Engineering Heat Transfer

J. R. Simonson
Senior Lecturer
The City University, London

SECOND EDITION

MACMILLAN
EDUCATION

First edition 1975
Reprinted 1978, 1981, 1983, 1984
Second edition 1988

Published by
MACMILLAN EDUCATION LTD
Houndmills, Basingstoke, Hampshire RG21 2XS
and London
Companies and representatives
throughout the world

Printed in Hong Kong

British Library Cataloguing in Publication Data
Simonson, J. R.
  Engineering heat transfer.—2nd ed.
  1. Heat—Transmission
  I. Title
  621.402′2   TJ265
  ISBN 0–333–47407–4 (hardback)
  ISBN 0–333–45999–7 (paperback)

Contents

vii

Preface to First Edition

The aim of this book, which is a revised edition of a book previously published by McGraw-Hill, is to introduce the reader to the subject of heat transfer. It will take him sufficiently along the road to enable him to start reading profitably the many more extensive texts on the subject, and the latest research papers to be found in scientific periodicals. This book is therefore intended for students of engineering in universities and technical colleges, and it will also be of assistance to the practising engineer who needs a concise reference to the fundamental principles of the subject. The engineering student will find most, if not all, aspects of the subject taught in undergraduate courses and, thus equipped, he will be in a position to undertake further studies at postgraduate level.

The aim throughout has been to introduce the principles of heat transfer in simple and logical steps. The need for an easily assimilated introduction to a subject becomes more urgent when the subject itself continues to grow at an ever-increasing rate. It is hoped that the material selected and presented will be of value at all levels of readership. Indebtedness is acknowledged to all those, past and present, who have contributed to the science of heat transfer with their original work, and as far as possible detailed references are given at the end of each chapter. Also grateful thanks are extended to various persons and organizations for permission to use certain diagrams, tables, and photographs; credit for these is given at appropriate points throughout the text.

It is also hoped that in this edition the changes made will further enhance the value of the book. Greater attention has been given to numerical methods in conduction, and some basic procedures in digital computing are included. The chapter on radiation has been extended to include an introduction to non-luminous gas radiation and a short section on solar radiation. Numerous small changes have been made throughout in the light of reviews and criticisms received. New worked examples are included to extend the range of applicability, and some of the original problems set have been replaced by more recent ones. SI units are now used exclusively, and conversion factors for British units are included in appendix 2.

Many of the problems included are university examination questions; the source is stated in each case. Where necessary the units in the numerical examples have been converted to SI. Indebtedness is acknowledged to the owners of the copyright of these questions for permission to use them, and for permission to convert the units. The universities concerned are in no way committed to the approval of numerical answers quoted.

Much of the material in this book has been taught for a number of years at undergraduate level to students at The City University. Grateful thanks are due to Professor J. C. Levy, Head of the Department of Mechanical Engineering, and to Mr B. M. Hayward, Head of the Thermodynamics Section. Discussions with colleagues at City and elsewhere have also contributed in numerous ways, and for this help sincere thanks are expressed.

Finally, thanks are due to Malcom Stewart, of The Macmillan Press, who has been responsible for the production of both editions, and also to my wife, who has typed the manuscript revisions.

Department of Mechanical Engineering,   JOHN R. SIMONSON
The City University

Preface to Second Edition

The essential aims of this new edition remain unchanged. While the subject matter of heat transfer at undergraduate level has not greatly altered, the student now has a powerful computational tool available to him, which in its use enables him to plot, explore and appreciate the mechanisms of heat transfer and their contributions in solving engineering problems. The use of the personal computer is increasing in all subject areas of undergraduate courses; and with growing emphasis on Design as an essential concept in the presentation of undergraduate studies, the introduction of computing methods into this edition forms the essential new material.

Relatively elementary computing procedures may be introduced in the subject matter of steady state and transient conduction, extended surfaces and heat exchangers, and the bulk of the new material lies in these areas. Since a considerable amount of valuable and relatively simple computing practice is possible in the field of cross flow heat exchange and in rotary regenerators, new sections have been added in these areas, and in order to make room for all the new computing material the chapter on mass transfer has been removed. More advanced computing techniques arise in convection studies and this subject is well covered in the literature. It is hoped that this new edition will help the student become familiar with the possibilities of computer literacy in the more elementary aspects of the subject of heat transfer.

The language of the computer listings is BASIC, which is the most popular language in use in the programming of personal and micro-computers. Some minor editing may be required to enable the given listings to run on particular machines. No claim is made for elegance in the programming presented; it is intended merely to present relatively simple examples with which the majority of students may gain in computing experience.

Some of the older problems have been removed and new ones introduced. At the same time, some earlier misprints and one or two misconceptions have been rectified. The author is grateful for comments and suggestions received since the first edition appeared, and he is grateful, too, for the support received which has made this second edition possible.

JOHN R. SIMONSON

Nomenclature

a	distance increment
A	area
b, l, t, w	linear dimension
C	capacity ratio of heat exchanger
C, K	constants of integration
Cd	average friction factor
Cf	skin friction coefficient
c_p	specific heat at constant pressure
C_p	volumetric specific heat at constant pressure
d	diameter
E	effectiveness of heat exchanger
f	friction factor
F	geometric configuration factor
$\mathscr{F}$	geometric emissivity factor
f_D	drag factor
g	gravitational acceleration
G	irradiation, mass velocity
Gz	Graetz number, $Re\ Pr(d/x)$
h_R	convection coefficient
H	product hA
h_{fg}	latent enthalpy of evaporation
h_r	radiation coefficient
i	current density
I	current
I	intensity of radiation
J	radiosity
k	thermal conductivity
L, D, T, W	linear dimension
L, M, T, θ	dimensions of length, mass, time, temperature
m	mass flow, or mass in transient conduction
$\dot{m}$	mass flow, where a non-flow m also occurs
n	coordinate direction
n	frequency of temperature variation
NTU	number of transfer units
$p, P, \Delta p$	pressure, difference of pressure
P	perimeter
PN	plate number
q	heat transfer per unit area and time
q'	heat generation per unit volume and time
Q	heat transfer per unit time, or a physical variable in dimensionless analysis

r	radius, radial direction
r	residual value
R	resistance
R_m	universal gas constant
S	scaling factors in electrical analogy
S_i	electrical shape factor
S_q	thermal shape factor
t	temperature
T	absolute temperature
$t, \Delta t, T$	time, time increment, time constant
U, U_A, U_L	overall heat transfer coefficients
U	velocity of temperature wave
v	velocity
v	specific volume
V	electrical potential, volume
x, y, z	coordinate direction, linear dimension
X	length of temperature wave
α	thermal diffusivity
α	absorptivity
β	coefficient of cubical expansion
δ	boundary layer thickness
δ_b	thickness of laminar sub-boundary layer
δ_t	thermal boundary layer thickness
δ_t'	equivalent conducting film thickness
ε	emissivity
ε	eddy diffusivity
ε_q	eddy thermal diffusivity
η_f	fin effectiveness
η_{fe}	equivalent effectiveness of finned surface
θ, θ_m	temperature difference, logarithmic temperature difference
θ	angle in cylindrical coordinate system
λ	wave-length
μ	dynamic viscosity
v	kinematic viscosity
ρ	density
ρ	electrical resistivity
ρ	reflectivity
σ	Stefan–Boltzmann constant, surface tension
τ	shear stress
τ	transmissivity
τ_t	turbulent shear stress
ϕ	angle in spherical coordinate system

Dimensionless groups

F	Fourier number, $\Delta t \alpha / a^2$
Gr	Grashof number, $\beta g \theta \rho^2 l^3 / \mu^2$
J	Colburn J-factor, $St.\ Pr^{2/3}$
Nu_l	Nusselt number, hl/k

Pr	Prandtl number, $c_p \mu/k$
Ra	Rayleigh number, $Gr.Pr$
Re_l	Reynolds number, $\rho v l/\mu$
St	Stanton number, $h/\rho v c_p$

Suffices

a	at axis of tube
b	black body
b	limit of laminar sub-boundary layer
c	cold fluid
C	convection
d, l, x	length terms used in dimensionless groups
e	equivalent
f	fluid
h	hot fluid, heated length
i, o	inlet, outlet (in heat exchangers)
l	liquid
m	mean value
M	metal, in heat-exchanger wall
n	direction of component
O	datum length
p	constant pressure
r	radial direction, or radial position
R	radiation
s	surroundings, of free stream
sat.	saturated temperature
t	temperature, turbulent
v	constant volume, vapour
w	wall
x, y, z	direction of component
θ	angular component
λ	monochromatic

Superscript

—	average value

1

Introduction

One of the primary concerns of the engineer is the design and construction of machines many times more powerful than himself or any of his domestic animals. The development of this skill over the centuries has been fundamental to the growth of civilization. Man's early efforts to harness the power of wind and water owed very little to engineering science, and indeed the early steam engine was a practical reality *before* the science of thermodynamics was firmly established. In contrast, there is now a vast fund of engineering knowledge behind the present day prime movers.

Much engineering activity is directed to the controlled release of power from fossil and nuclear fuels, and with making that power available where it is needed. The laws of heat transfer are of the utmost importance in these activities. The generation of power from the energy changes of chemical and nuclear reactions involves the transfer of vast quantities of thermal energy. Further, chemical processes of combustion yield temperatures at which most constructional materials would melt; adequate protection by heat transfer processes is therefore vital. The distribution of energy as electricity is accompanied, at all stages, by certain wastages manifested as rising temperature of the equipment. Heat transfer considerations enable these temperatures to be controlled within safe limits.

The laws of heat transfer find application in many other fields of engineering. Chemical and process engineering, and manufacturing and metallurgical industries are examples. In addition, the civil and constructional engineer and environment control engineer need considerable knowledge of the subject. Large city buildings must be economically heated and insulated, and air conditioning is increasingly necessary.

To the mechanical engineer heat transfer is a subject closely allied to applied thermodynamics. The first and second laws of

1

thermodynamics state the relations between the physical entities of heat and work, and the limit to the amount of work that may be obtained from any source of heat. Even this limit cannot be reached in practical engineering processes because of their inherent irreversibility. These irreversibilities may be accounted for in calculations but, even so, thermodynamics alone leaves a lot of questions unanswered. There is no time scale and, consequently, thermodynamics will not permit the calculation of physical sizes necessary to achieve a given objective. In a steam power plant it is necessary to transfer the thermal energy of the hot combustion gases of the burnt fuel to the water in the boiler tubes. The actual rate of transfer to produce a required flow rate of steam may be known, but without the laws of heat transfer and knowledge of the properties of the engineering materials to be used, it is not possible to calculate the size and surface area of the tubes required. From an economic point of view, the boiler must be made as small as possible, hence the heat transfer rate must be as high as possible. Elsewhere in the plant, heat transfer considerations are necessary in insulating the steam delivery lines and in condensing the low pressure turbine exhaust.

Heat transfer processes, then, are described by equations which relate the energy to be transferred in unit time to the physical area involved. Other factors entering the equations are the temperatures, or the temperature gradient, and some coefficient which depends on various physical properties of the system and on the particular mechanism of heat transfer involved. Three basic mechanisms of heat transfer are recognized. They may occur separately, or simultaneously. Separate equations may be written to describe each mechanism, and when two or more mechanisms occur simultaneously it is sometimes possible to add the separate effects; but sometimes it is necessary to consider the equations of the participating mechanisms together. The subject matter thus conveniently sub-divides itself into the separate basic mechanisms of heat transfer, and the combinations of them.

Heat is transferred by conduction, convection, and radiation. Before describing these processes, it is desirable to clarify what is meant by 'heat'. In the study of thermodynamics, heat is defined as an energy transfer between communicating systems, arising solely from a temperature difference. Thus a heat transfer is strictly a phenomenon occurring only at *boundaries* of systems, and a heat transfer elsewhere in a system is more correctly a redistribution of internal energy within the system. As it is convenient to keep to the conventional language of heat transfer, this should be kept in mind, and the word heat will not in most cases be in accord with the thermodynamic usage.

Conduction is the mode of heat transfer in a solid material and occurs by virtue of a temperature difference between different parts of the material. Conduction also occurs in liquids and gases but is generally associated also with convection, and possibly with radiation as well in the case of gases. Conduction within a solid is a transfer of internal energy; this energy is, in fact, energy of motion of the constituent molecules, atoms, and particles of which the material consists. The kinetic energy of the motion is proportional to the absolute temperature; molecular collisions lead to energy transfer to regions of lower kinetic energy. Under steady conditions a molecule will pass on the same amount of energy that it receives. Under non-steady conditions the flow of energy is governed by the changing energy levels.

The theory of conduction heat transfer was established by Joseph Fourier whose work was published in Paris in 1822,[1] but pioneer work was done by Biot in 1804[2] and 1816.[3] Conduction is described by an equation known as the Fourier rate equation

$$Q_x = -kA\frac{dt}{dx} \qquad (1.1)$$

The rate of heat flow (in only the x-direction, see Fig. 1.1) is proportional to the product of the area of flow and the temperature gradient, the constant of proportionality being the thermal conductivity k which is a property of the material. The negative sign results from the convention of defining a positive heat flow in the direction of a negative temperature gradient. The property k may be a function of temperature and direction of heat flow. Materials with directional dependence of thermal conductivity are said to be anisotropic.

The units involved depend on the system chosen. In the SI system, the unit of heat or internal energy is the joule, hence rate of heat transfer is measured in J/s or W. However, the kilojoule, (kJ), and kilowatt, (kW), are accepted multiples of the SI unit, and to be consistent with general usage in thermodynamics, the kJ and kW are the preferred units in this book. With the area in m^2 and the temperature gradient in K/m, the units of k are kW/(m K). This follows the British Standards recommendation[4] for the presentation of complex units.

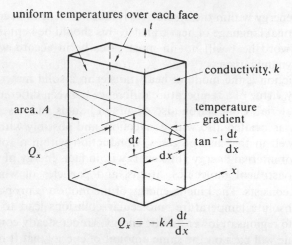

$$Q_x = -kA\frac{dt}{dx}$$

Fig. 1.1. Fourier's law for one-dimensional steady state conduction in a plane slab of material.

In the British system of units, used in the majority of publications in English up to the mid 1960s, the rate of heat transfer is measured in British thermal units/hour, or Btu/h, and with the area in ft² and the temperature gradient in °F/ft, k is measured in Btu/(ft h °F).

Conduction in fluids generally forms a very small part of the total heat transfer, convection being the predominating mechanism. Convection is the name given to the gross motion of the fluid itself, so that fresh fluid is continually available for heating or cooling. Apart from the bulk movement of the fluid, there is generally a smaller motion of eddies which further assists in distributing heat energy. Convection heat transfer is sub-divided into two different kinds, natural and forced. Heat transfer by *natural convection* occurs between a solid and a fluid undisturbed by other effects when there is a temperature difference between the two, as in a kettle of water. It is not often that a fluid can be regarded as entirely at rest, so frequently there is a small amount of forced convection as well. But true *forced convection* requires a major applied motion of the fluid in relation to the source or sink of heat, so that natural convection effects are negligible. An important aspect of natural convection is that the fluid motion which does occur is due entirely to natural buoyancy forces arising from a changing density of the fluid in the vicinity of the surface. Within the realms of both natural and forced convection there are two sub-divisions of laminar and turbulent flow convection. In forced convection separation of flow can occur from the surfaces of immersed bodies, for example, in the flow across the outside of a pipe. A further type of forced convection involves a phase change of the fluid, as in boiling and condensing.

It is thus evident that many factors enter into heat convection, including the shape and magnitude of the solid–fluid boundary, characteristics of the fluid flow, such as the magnitude of turbulent eddies, and the conductivity of the fluid itself.

Because of these complexities many convection problems are not amenable to mathematical solution, and recourse is made to techniques of dimensional analysis and experiment. Thus many empirical dimensionless relationships are now available in the literature to enable the engineer to design his heat transfer apparatus, whether it be an industrial heat exchanger or domestic convector.

Newton (1701)[5] proposed a general equation to describe convection heat transfer.

$$Q = hA(t_1 - t_2) \tag{1.2}$$

Figure 1.2 indicates that heat transfer is occurring from a surface of area A at temperature t_1 to a fluid at a lower temperature t_2.

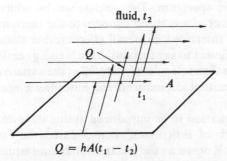

$$Q = hA(t_1 - t_2)$$

Fig. 1.2. Diagrammatic representation of convection from a flat plate, illustrating the use of Newton's equation.

h is the convection coefficient and has the units $kW/(m^2\ K)$ in the SI system, or $Btu/(ft^2\ h\ °F)$ in the British system. It takes care of the many factors entering a particular example of convection, and the value of h will vary for differing flow regimes, fluid properties, and temperature differences. The main problem in the analysis of

Table 1.1

Convection system	Range of h,* kW/(m² K)
Natural convection	0·004–0·05
Forced convection (air)	0·01–0·55
Forced convection (liquids)	0·1–5·5
Boiling heat transfer (water)	1·0–110·0
Condensation (steam, filmwise)	0·55–25·0
Forced convection (liquid metals)	3·0–110·0

* For numerical conversion factors, see the Appendix

convection is to predict values of h for design purposes. The value of h in different regimes and for different fluids is generally within the ranges indicated in Table 1.1.

The third mode of heat transfer known as radiation is rather different in nature from the first two. Conduction and convection occur within solid or fluid material and often are present simultaneously. In contrast, radiation is an energy transfer which is transmitted most freely in a vacuum. It occurs between all material phases. All matter at temperatures above absolute zero emits electromagnetic waves of various wave-lengths. Visible light together with infra-red and ultra-violet radiation forms but a small part of the total electromagnetic spectrum. The mechanism by which radiation is propagated is not of any direct concern to the mechanical engineer, who is mostly interested in overall effects rather than in molecular detail. It is sufficient to say that radiation is energy emitted by vibrating electrons in the molecules of material at the surface of a body, and the amount emitted depends on the absolute temperature of the body.

The third equation to be introduced at this stage dates from 1884 when the work of Boltzmann[6] consolidated the earlier work of Stefan (1879).[7] Known as the Stefan–Boltzmann equation, it is

$$Q = \sigma A T^4 \tag{1.3}$$

where T is the absolute temperature, A is the surface area of a perfectly radiating body and σ is the Stefan–Boltzmann constant and has the value of $56\cdot7 \times 10^{-12}$ kW/(m² K⁴), or $0\cdot171 \times 10^{-8}$ Btu/ (ft² h °R⁴). Stefan established this relationship experimentally, subsequently Boltzmann proved it theoretically. A perfectly radiating or black body emits at any given temperature the maximum possible energy at all wave-lengths. The energy emitted will be less for real materials. This equation defines an energy emission, rather than an energy exchange. The area A will also be absorbing radiation from elsewhere, which must be taken into account in an energy exchange relationship. The emitting and absorbing characteristics of surfaces, and the 'view' that surfaces have of each other, are factors which enter the consideration of radiation exchanges.

REFERENCES

1. Fourier, J. B. *Théorie analytique de la chaleur*, Paris, 1822. Translated by A. Freeman, Dover Publications, New York, 1955.
2. Biot, J. B. *Bibliothèque Britannique*, Vol. 27, 310 (1804).
3. Biot, J. B. *Traité de physique*, Vol. 4, 669 (1816).
4. British Standards Institution, *The Use of SI Units*, PD5686: 1972.
5. Newton, I. *Phil. Trans., Roy. Soc.*, London, Vol. 22, 824 (1701).
6. Boltzmann, L. *Wiedemanns Annalen*, Vol. 22, 291 (1884).
7. Stefan, J. *Sitzungsber. Akad. Wiss. Wien. Math.-naturw. Kl.*, Vol. 79, 391 (1879).

2

The equations of heat conduction

2.1 The Nature of Heat Conduction

The Fourier equation of heat conduction (1.1) has already been introduced. This equation is for one-dimensional heat flow, and may be written in a more general form:

$$Q_n = -kA\frac{\partial t}{\partial n} \qquad (2.1)$$

where Q_n is the rate of heat conduction in the n-direction, and $\partial t/\partial n$ is the temperature gradient in that direction. The partial derivative is used since there may exist temperature gradients in other directions. One-dimensional conduction does not often occur in practice since a body would have to be either perfectly insulated at its edges or so large that conduction would be one-dimensional at the centre.

Equation (2.1) expresses an instantaneous rate of heat transfer. It may be re-written

$$q_n = \frac{Q_n}{A} = -k\frac{\partial t}{\partial n} \qquad (2.2)$$

where q_n is the heat flux in heat units per unit time and per unit area in the n-direction. This is a vector quantity since it has magnitude and direction. The greatest heat flux at an isothermal surface will always occur along the normal to that surface.

Heat conduction within a solid may be visualized as a heat flux which varies with direction and position throughout the material. This follows from the fact that temperature within the solid is a function of position coordinates of the system (e.g., x, y, z). In addition, temperature may be a function of time, (t), so in general $t = f(x, y, z, t)$.

The problem of determining the magnitude of heat conduction

8

resolves itself to finding first the isotherms within the system and the way in which their positions vary with time. In steady state conduction the isotherms remain stationary with time, and one may visualize a large number of isothermal surfaces throughout the system, differing incrementally in temperature. The heat flux normal to any one surface will vary with position depending on the distance between surfaces. It is then necessary to sum the heat flow through the *boundary* surfaces if internal heat sources are present, or, if not, through *any* isothermal surface. In unsteady conduction the problem is complicated by the fact that isothermal surfaces are no longer fixed, and the rate at which heat is being stored must be taken into account.

Before taking the first step, which is to develop the equation for temperature as a function of position and time, it is opportune to introduce some facts about different conducting materials.

Solid materials may be divided into two groups, metallic and non-metallic, for which there is a marked contrast in the values of conductivity. The Appendix lists properties for some of the more useful materials. The high values of conductivity of metals are attributable to the well ordered crystalline structure of the material. The close arrangement of molecules permits a rapid transfer of energy and, in addition, free electrons play a considerable part. Metals such as copper which are good electrical conductors also conduct heat well. There is also a marked similarity between conduction heat transfer and the flow of electricity, and the electrical analogy is often used in the solution of conduction problems.

In contrast, non-metals do not have a well ordered crystalline structure and, in addition, are often porous in nature. Thus energy transfer between molecules is seriously impeded, and the values of conductivity are much lower. The small pores within the material, being full of air, further restrict the flow of heat since gases are poor conductors. This is because molecules of a gas are relatively widely spaced and the transfer of energy depends on collisions between these molecules.

The thermal conductivities of most substances vary with temperature, and for accuracy such variation should be allowed for in conduction problems. However, this is a complication which may be ignored in an introductory study of the subject because the variation with temperature is not great. Over a reasonable temperature range the relationship between conductivity k_t and temperature t may be assumed linear:

$$k_t = k_0(1 + \alpha t) \qquad (2.3)$$

where k_0 is the conductivity at temperature t_0, and α is a constant. In most practical applications it is sufficient to assume a mean uniform value for conductivity.

A complication more serious than temperature variation of conductivity occurs in certain engineering materials, viz., that conductivity may vary with the direction of heat flow. This arises commonly in laminated materials used in electrical engineering. Thus the conductivity parallel to the laminates is different to the value perpendicular to the laminates. Most types of wood also exhibit this property, the conductivity parallel to the grain being different to that across the grain. Conducting materials exhibiting this property are said to be *anisotropic*. In the absence of this property the material is said to be *isotropic*. For anisotropic materials the analysis of conduction is more difficult and is not included in this introductory text. The basic ideas are given by Eckert and Drake,[1] and the general treatment may be found in the work of Carslaw and Jaeger.[2]

Differential equations of the temperature field will now be developed in two coordinate systems, Cartesian and cylindrical.

2.2 The Differential Equation of Conduction in a Cartesian Coordinate System

The material of the system is assumed to be isotropic and the conductivity is assumed invariable with temperature. Consider the infinitesimal element of the material represented by the volume $dx\,dy\,dz$ in Fig. 2.1. The heat flowing into and out of the element is resolved in the x-, y- and z-directions. Thus from the Fourier equation the rate of heat flowing into the element in the x-direction is

$$dQ_x = -k\,dy\,dz\,\frac{\partial t}{\partial x}$$

since the area of flow normal to the x-direction is $dy\,dz$ and the temperature gradient is $\partial t/\partial x$. The rate of heat flowing out of the element in the x-direction is

$$dQ_{(x+dx)} = -k\,dy\,dz\,\frac{\partial}{\partial x}\left(t + \frac{\partial t}{\partial x}dx\right)$$

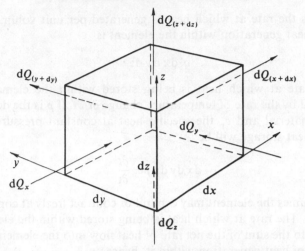

Fig. 2.1. Conduction in an element of material in Cartesian coordinates.

$$= - k \, dy \, dz \frac{\partial t}{\partial x} - k \, dx \, dy \, dz \frac{\partial^2 t}{\partial x^2}$$

Therefore the net rate of heat flow into the element in the x-direction is

$$dQ_x - dQ_{(x+dx)} = k \, dx \, dy \, dz \frac{\partial^2 t}{\partial x^2} \qquad (2.4)$$

In a similar manner, the net rates of heat flow into the element in the y- and z-directions are given by

$$dQ_y - dQ_{(y+dy)} = k \, dx \, dy \, dz \frac{\partial^2 t}{\partial y^2} \qquad (2.5)$$

$$dQ_z - dQ_{(z+dz)} = k \, dx \, dy \, dz \frac{\partial^2 t}{\partial z^2} \qquad (2.6)$$

The total rate of heat flow into the element is the sum of the right-hand sides of equations (2.4), (2.5), and (2.6), which is

$$k \, dx \, dy \, dz \left(\frac{\partial^2 t}{\partial x^2} + \frac{\partial^2 t}{\partial y^2} + \frac{\partial^2 t}{\partial z^2} \right)$$

In addition to heat flowing into and out of the element, the possibilities of heat being generated within the element (e.g., due to the flow of electricity) and of heat being stored within the element (as in the case of unsteady conduction) have to be considered.

If q' is the rate at which heat is generated per unit volume, the rate of heat generation within the element is

$$q' \, dx \, dy \, dz$$

The rate at which heat is being stored within the element is governed by the rate of temperature change $\partial t/\partial t$. If ρ is the density of the material and c_p the specific heat at constant pressure, the rate of heat storage will be

$$dx \, dy \, dz \, \rho c_p \frac{\partial t}{\partial t}$$

This assumes the element may expand or contract freely at constant pressure. The rate at which heat is being stored within the element is equal to the sum of the net rate of heat flow into the element and the rate of heat generation within it, hence:

$$\rho c_p \frac{\partial t}{\partial t} = k\left(\frac{\partial^2 t}{\partial x^2} + \frac{\partial^2 t}{\partial y^2} + \frac{\partial^2 t}{\partial z^2}\right) + q'$$

$$\therefore \qquad \frac{\partial t}{\partial t} = \alpha\left(\frac{\partial^2 t}{\partial x^2} + \frac{\partial^2 t}{\partial y^2} + \frac{\partial^2 t}{\partial z^2}\right) + \frac{q'}{\rho c_p} \qquad (2.7)$$

where $\alpha = k/\rho c_p$ and is known as the thermal diffusivity of the material. It is a ratio of the heat conduction to heat storage qualities of the material.

Equation (2.7) is the general differential equation of conduction in a Cartesian coordinate system and may be simplified to suit any particular application. Thus the equation for unsteady conduction in one dimension without heat generation is

$$\frac{\partial t}{\partial t} = \alpha\left(\frac{\partial^2 t}{\partial x^2}\right) \qquad (2.8)$$

since q', $\partial^2 t/\partial y^2$ and $\partial^2 t/\partial z^2$ are equal to 0.

For any problem of steady conduction, $\partial t/\partial t = 0$, since there is then no variation of temperature with time. The equations for two- and one-dimensional steady conduction with heat generation are

$$0 = \alpha\left(\frac{\partial^2 t}{\partial x^2} + \frac{\partial^2 t}{\partial y^2}\right) + \frac{q'}{\rho c_p} \qquad (2.9)$$

and

$$0 = \alpha\left(\frac{d^2 t}{dx^2}\right) + \frac{q'}{\rho c_p} \tag{2.10}$$

it being permissible to use the total derivative in the one-dimensional case. In the absence of heat generation the equations reduce to

$$0 = \alpha\left(\frac{\partial^2 t}{\partial x^2} + \frac{\partial^2 t}{\partial y^2}\right) \quad \text{and} \quad 0 = \alpha\left(\frac{d^2 t}{dx^2}\right)$$

and, consequently,

$$\frac{\partial^2 t}{\partial x^2} + \frac{\partial^2 t}{\partial y^2} = 0 \tag{2.11}$$

and

$$\frac{d^2 t}{dx^2} = 0 \tag{2.12}$$

Problems involving equations (2.8) to (2.12) will be considered in later chapters.

2.3 The Differential Equation of Conduction in a Cylindrical Coordinate System

Often, conduction problems involve heat flow in solid or hollow round bars and, consequently, the cylindrical coordinate system, Fig. 2.2, is used. The general approach is exactly the same as before except that heat flows in radial, circumferential, and axial directions have now to be considered. The element to be considered has volume $r d\theta \, dr \, dz$. Heat flowing into the element in the radial direction is

$$dQ_r = -k \, dz \, r d\theta \frac{\partial t}{\partial r}$$

and out of the element in the radial direction,

$$dQ_{(r+dr)} = -k \, dz(r + dr) \, d\theta \frac{\partial}{\partial r}\left(t + \frac{\partial t}{\partial r} dr\right)$$

Hence

$$dQ_r - dQ_{(r+dr)} = k \, dz \, dr \, d\theta \frac{\partial t}{\partial r} + k \, dz \, r d\theta \frac{\partial^2 t}{\partial r^2} dr \tag{2.13}$$

neglecting a term of higher order.

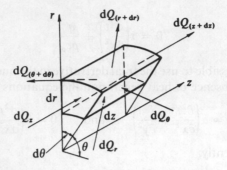

Fig. 2.2. Conduction in an element of material in cylindrical coordinates.

In a similar manner, the net heat flowing into the element in the circumferential direction is found to be

$$dQ_\theta - dQ_{(\theta + d\theta)} = k \, dr \, dz \frac{\partial^2 t}{r^2 \partial \theta^2} r d\theta \qquad (2.14)$$

and, in the axial direction,

$$dQ_z - dQ_{(z + dz)} = k \, dr \, rd\theta \frac{\partial^2 t}{\partial z^2} dz \qquad (2.15)$$

The rate of heat generation within the element is

$$q' \, rd\theta \, dr \, dz$$

and the rate at which heat is being stored within the element is

$$rd\theta \, dr \, dz \, \rho c_p \frac{\partial t}{\partial t}$$

Then an energy balance for the element leads to the general differential equation for heat flow in three dimensions in a cylindrical coordinate system, i.e.,

$$\frac{\partial t}{\partial t} = \alpha \left(\frac{\partial^2 t}{\partial r^2} + \frac{1}{r} \frac{\partial t}{\partial r} + \frac{1}{r^2} \frac{\partial^2 t}{\partial \theta^2} + \frac{\partial^2 t}{\partial z^2} \right) + \frac{q'}{\rho c_p} \qquad (2.16)$$

This equation may similarly be simplified to suit any particular problem. Steady one-dimensional heat flow in the radial direction only will be considered in later chapters. With heat generation within the material the equation is

$$\alpha\left(\frac{d^2 t}{dr^2} + \frac{1}{r}\frac{dt}{dr}\right) + \frac{q'}{\rho c_p} = 0 \tag{2.17}$$

and without heat generation,

$$\frac{d^2 t}{dr^2} + \frac{1}{r}\frac{dt}{dr} = 0 \tag{2.18}$$

PROBLEM

Show that the general equation of heat conduction in spherical coordinates is given by

$$\frac{\partial t}{\partial t} = \alpha\left[\frac{\partial^2 t}{\partial r^2} + \frac{2}{r}\frac{\partial t}{\partial r} + \frac{1}{r^2 \sin\phi}\frac{\partial}{\partial\phi}\left(\sin\phi\frac{\partial t}{\partial\phi}\right) + \frac{1}{r^2 \sin^2\phi}\frac{\partial^2 t}{\partial\theta^2}\right] + \frac{q'}{\rho c_p}$$

and transform the equation in rectangular coordinates (2.7) into spherical coordinates by making the substitutions

$$x = r\sin\phi\cos\theta$$
$$y = r\sin\phi\sin\theta$$
$$z = r\cos\phi$$

REFERENCES

1. Eckert, E. R. G., and Drake, R. M. *Introduction to the Transfer of Heat and Mass*, 2nd ed., McGraw-Hill, New York (1959).
2. Carslaw, H. S., and Jaeger, J.C. *Conduction of Heat in Solids*, Oxford University Press (1947).

3

One-dimensional steady state conduction

The simplest example of steady state conduction in one dimension is the transfer of heat through a single plane slab. Many simple problems, such as conduction through the wall of a building, approximate to this.

3.1 Conduction in Plane Slabs

To calculate the conduction rate in a single slab of isotropic invariable thermal conductivity material, Fourier's law applied to an infinitesimal layer within the slab, Fig. 3.1, may be integrated directly. Thus

$$Q_x = - kA \frac{dt}{dx}$$

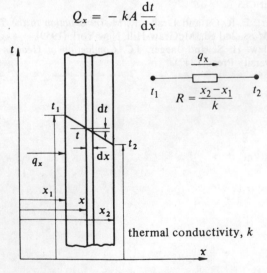

Fig. 3.1. One-dimensional steady state conduction in a plane slab.

16

and hence

$$Q_x = \frac{-kA(t_2 - t_1)}{(x_2 - x_1)} \qquad (3.1)$$

where $(x_2 - x_1)$ is the thickness of the slab and A is the area of the slab. Using consistent units, the heat transfer calculated will be in heat units per unit time.

The same result will be obtained if the appropriate differential equation is integrated. Integration twice of equation (2.12),

$$\frac{d^2t}{dx^2} = 0$$

gives

$$t = C_1 x + C_2 \qquad (3.2)$$

where C_1 and C_2 are constants of integration to be determined from the boundary conditions, i.e., the temperatures at x_1 and x_2. Equation (3.2) indicates that the temperature variation through the slab is linear. The temperature gradient from equation (3.2) used in Fourier's law gives equation (3.1). Equation (3.1) may be re-written as

$$\frac{Q_x}{A} = q_x = \frac{k(t_1 - t_2)}{x_2 - x_1}, \qquad (3.3)$$

in which form it may be compared with Ohm's law describing the flow of electricity, i.e.,

$$\text{Current density } (i) = \frac{(V_1 - V_2)}{\rho(x_2 - x_1)}, \qquad \frac{\text{Potential difference}}{\text{Resistance of unit area}}$$

where ρ is the resistivity of the material, in units of ohms $\times$ length. The heat flux q_x is analogous to current density i; the temperature drop $(t_1 - t_2)$ is analogous to potential difference $(V_1 - V_2)$; and the resistance per unit area to heat transfer, $(x_2 - x_1)/k$, is analogous to electrical resistance per unit area, $\rho(x_2 - x_1)$. The usefulness of this similarity will be made more apparent later.

Conduction through a system of plane slabs of different material has often to be considered. A partition wall comprising two layers of plaster board separated by a thickness of glass-fibre insulation, or a furnace wall consisting of a layer of fire brick and a layer of insulating brick, are typical examples. Further, such a system may separate two fluids of different temperatures, when the actual wall temperatures are not known. The processes of heat transfer between the wall surfaces and the adjacent fluid are by convection and radiation. Figure 3.2 shows such a system. The Newton equation for convection may be written in the sign convention of equation (3.1). Thus

$$q_c = -h_c(t_f - t_w) \tag{3.4}$$

In this equation, q_c is the heat flux due to convection at the solid/fluid interface, and t_w is the wall temperature and t_f the fluid temperature. The region in the fluid where the temperature changes from t_f to t_w is known as the boundary layer. h_c is the convection coefficient and is assumed known. Its determination forms the subject matter of later chapters, where the suffix c is dropped.

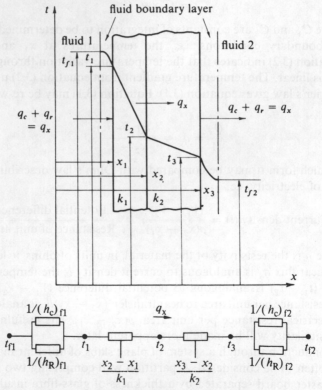

Fig. 3.2. *A multiple plane slab separating two fluids, one-dimensional steady state conduction.*

As a convenience, the radiation exchange between the wall and fluid or some other surface beyond the fluid may be expressed by an analogous equation

$$q_r = - h_R(t_f - t_w) \tag{3.5}$$

Since radiation exchanges are a function of the fourth power of the absolute temperatures involved, the radiation coefficient h_R is heavily temperature-dependent.

The total heat flow or conduction flux q_x from the wall by convection and radiation is found by adding (3.4) and (3.5):

$$q_x = q_c + q_r = - h_c(t_f - t_w) - h_R(t_f - t_w)$$

$$= - (h_c + h_R)(t_f - t_w) \tag{3.6}$$

A multiple slab of two layers of conductivities k_1 and k_2 which separates two fluids f_1 and f_2 at temperatures t_{f1} and t_{f2} is now considered. For exchange between fluid f_1 and wall surface at t_1:

$$q_x = - (h_c + h_R)_{f1}(t_1 - t_{f1}) \tag{3.7}$$

For conduction through the two layers of material:

$$q_x = \frac{- k_1(t_2 - t_1)}{x_2 - x_1} = \frac{- k_2(t_3 - t_2)}{x_3 - x_2} \tag{3.8}$$

For exchange between the wall surface at t_3 and the fluid f_2:

$$q_x = - (h_c + h_R)_{f2}(t_{f2} - t_3) \tag{3.9}$$

Re-arranging and adding:

$$q_x \left[\frac{1}{(h_c + h_R)_{f1}} + \frac{x_2 - x_1}{k_1} + \frac{x_3 - x_2}{k_2} + \frac{1}{(h_c + h_R)_{f2}} \right]$$

$$= - (t_{f2} - t_{f1})$$

and hence

$$q_x = - U(t_{f2} - t_{f1}) \tag{3.10}$$

where

$$\frac{1}{U} = \frac{1}{(h_c + h_R)_{f1}} + \frac{x_2 - x_1}{k_1} + \frac{x_3 - x_2}{k_2} + \frac{1}{(h_c + h_R)_{f2}} \tag{3.11}$$

$1/U$ is the overall thermal resistance per unit area between fluids and U is the overall heat transfer coefficient. The resistances to heat flow due to convection and radiation act in parallel and the resistances due to the conducting layers act in series. The heat flow is calculated from (3.10), once the overall coefficient U is found from (3.11), and interface temperatures follow from (3.7), (3.8), and (3.9).

Sometimes in composite structures slabs of differing thermal conductivity are present as shown in Fig. 3.3. This situation may be treated one-dimensionally provided it is assumed that the y–z faces of the intermediate slabs have uniform temperatures. The total resistance may be deduced by adding the intermediate resistances in parallel before adding the others in series.

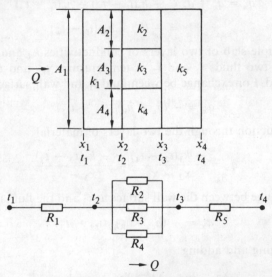

Fig. 3.3. One-dimensional steady state conduction in a series–parallel system of slabs; t_2 and t_3 are uniform temperatures in the y-z plane.

EXAMPLE 3.1

The reduction of heat loss from buildings is of very great practical and economic importance. The Chartered Institution of Building Services Engineers in the CIBS Guide Book A, give values of overall heat transfer coefficient (known as U values and expressed as W/(m^2 K) and not kW/(m^2 K)) for various types of wall, window, and roof. Some typical values are given in Appendix 3. Use the U values given below to calculate the heat transfer rate through a house structure in cases (i) and (ii). In both cases the wall area is $110 \, \text{m}^2$, wood-frame window total area $14 \, \text{m}^2$, upstairs ceiling area $36 \, \text{m}^2$, environment temperature difference 21°C.

Case (i) 335 mm solid brick wall, $U = 1.5 \, \text{W}/(\text{m}^2 \, \text{K})$; pitched roof with felt, foil-backed board ceiling, $U = 1.5 \, \text{W}/(\text{m}^2 \, \text{K})$; single-glazed windows, $U = 4.3 \, \text{W}/(\text{m}^2 \, \text{K})$.

Case (ii) 335 mm solid wall plus 30 mm foam board lining, $k = 0.026 \, \text{W}/(\text{m K})$; pitched roof as before plus 50 mm glass-fibre insulation $U = 0.5$; double-glazed windows, $U = 2.5 \, \text{W}/(\text{m}^2 \, \text{K})$.

Solution. For *parallel* heat flow through walls, windows and roof, in case (i)

$$Q = 21(110 \times 1.5 + 36 \times 1.5 + 14 \times 4.3) = 5860 \, \text{W}$$
$$= 5.86 \, \text{kW}$$

In case (ii), the thermal resistance of the insulated wall is the original resistance plus the insulation resistance which equals $1/1.5 + 0.03/0.026 = 1.82$.

New U value $= 1/1.82 = 0.55$

$$Q = 21(110 \times 0.55 + 36 \times 0.5 + 14 \times 2.5) = 2380 \, \text{W}$$
$$= 2.38 \, \text{kW}$$

A saving of $3.48 \, \text{kW}$ is achieved. Actual heating requirements will be greater than the figures calculated on account of air changes, and some losses through the ground floor.

EXAMPLE 3.2

Heat transfer through a double-glazed window is an example of the application of resistances in parallel and in series. A section of the window together with the corresponding diagram of circuit resistances is shown in Fig. 3.4. Convection and radiation coefficients act in parallel on both inside and outside surfaces of the frame and glazing, and the overall resistances of the frame and glazing are in parallel. The glazing itself consists of three resistances in series. For a small width, heat transfer across the air gap is by pure conduction. As the width increases, some convection commences so that an optimum width of air gap occurs. The convection effect is accounted for by an empirical term in the air gap resistance, R_5. The BASIC names of all variables may be deduced from the listing, with the exception of TKF, TKA and TKG which are the thermal conductivities of the frame, air gap and glass, respectively. The BASIC program for this problem is listed below; variables in the data are given in answer to questions appearing. A sample run is also shown. The variation of heat transfer as a function of air gap is given in the table:

Air gap, mm	Heat transfer, W
3	157·32
7	107·92
11	99·56
15	103·96
20	114·85
25	126·87

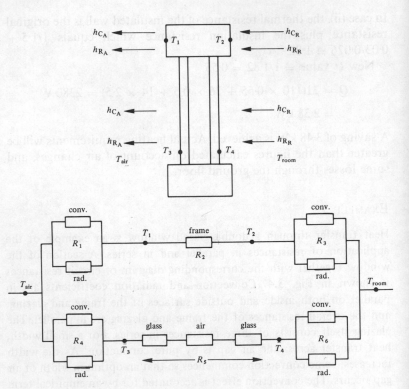

Fig. 3.4. *Diagrammatic representation of a double-glazed window and the equivalent thermal resistance circuit.*

BASIC Program Listing

```
10      PRINT,"PLEASE TYPE CONVECTION COEFFICIENT, AIR SIDE IN W/M2 K:"
20      INPUT HCA
30      PRINT,"PLEASE TYPE RADIATION COEFFICIENT, AIR SIDE IN W/M2 K:"
40      INPUT HPA
50      PRINT,"PLEASE TYPE CONVECTION COEFFICIENT, ROOM SIDE IN W/M2 K:"
60      INPUT HCR
70      PRINT,"PLEASE TYPE RADIATION COEFFICIENT, ROOM SIDE IN W/M2 K:"
80      INPUT HRR
90      PRINT,"PLEASE TYPE WINDOW FRAME AREA, IN M2:"
100     INPUT WFA
110     PRINT,"PLEASE TYPE WINDOW FRAME THICKNESS, IN M:"
120     INPUT WFT
130     PRINT,"PLEASE TYPE WINDOW GLASS AREA, IN M2:"
140     INPUT WGA
150     PRINT,"PLEASE TYPE WINDOW GLASS THICKNESS, IN MM:"
160     INPUT WGT
165     WGT=WGT/1000.0
170     PRINT,"PLEASE TYPE AIR GAP THICKNESS, IN MM:"
180     INPUT AGT
185     AGT=AGT/1000.0
190     TKF=0.166
195     TKA=0.026
200     TKG=0.762
205     PRINT,"PLEASE TYPE ROOM TEMPERATURE IN DEG C:"
210     INPUT TROOM
215     PRINT,"PLEASE TYPE EXTERNAL AIR TEMPERATURE IN DEG C:"
220     INPUT TAIR
230     R1=1.0/(WFA*(HCA+HRA))
240     R2=WFT/(WFA*TKF)
250     R3=1.0/(WFA*(HCR+HRR))
260     R4=1.0/(WGA*(HCA+HRA))
270     R5=2.0*WGT/(WGA*TKG)+1.0/(TKA*WGA/AGT+(AGT-0.007)*350.0)
280     R6=1.0/(WGA*(HCR+HRR))
290     RF=R1+R2+R3
300     RW=R4+R5+R6
310     ROV=RF*RW/(RW+RF)
320     QT=(TROOM-TAIR)/ROV
330     QF=(TROOM-TAIR)/RF
340     QW=(TROOM-TAIR)/RW
350     T1=TROOM-(R2+R3)*(TROOM-TAIR)/RF
360     T2=TROOM-R3*(TROOM-TAIR)/RF
370     T3=TROOM-(R5+R6)*(TROOM-TAIR)/RW
380     T4=TROOM-R6*(TROOM-TAIR)/RW
390     PRINT,"      RESULTS FOR AIR GAP THICKNESS:",AGT
400     PRINT,"TOTAL HEAT TRANSFER RATE IN W:",QT
410     PRINT,"HEAT TRANSFER RATE, FRAME, IN W:",QF
420     PRINT,"HEAT TRANSFER RATE, WINDOW, IN W:",QW
430     PRINT,"OUTSIDE GLASS TEMPERATURE, IN DEG C:",T3
440     PRINT,"INSIDE GLASS TEMPERATURE, IN DEG C:",T4
450     PRINT,"OUTSIDE FRAME TEMPERATURE, IN DEG C:",T1
460     PRINT,"INSIDE FRAME TEMPERATURE, IN DEG C:",T2
470     PRINT,"    DO YOU WISH TO DO A REPEAT CALCULATION? 1-YES,2-NO:"
480     INPUT IX%
490     IF(IX%<2)GO TO 10
500     STOP
```

Sample Result from Listing

```
PLEASE TYPE CONVECTION COEFFICIENT, AIR SIDE IN W/M2 K:
=12.0
PLEASE TYPE RADIATION COEFFICIENT, AIR SIDE IN W/M2 K:
=5.0
PLEASE TYPE CONVECTION COEFFICIENT, ROOM SIDE IN W/M2 K:
=4.0
PLEASE TYPE RADIATION COEFFICIENT, ROOM SIDE IN W/M2 K:
=9.0
PLEASE TYPE WINDOW FRAME AREA, IN M2:
=0.4
PLEASE TYPE WINDOW FRAME THICKNESS, IN M:
=0.12
PLEASE TYPE WINDOW GLASS AREA, IN M2:
=1.6
PLEASE TYPE WINDOW GLASS THICKNESS, IN MM:
=4.0
PLEASE TYPE AIR GAP THICKNESS, IN MM:
=25.0
PLEASE TYPE ROOM TEMPERATURE IN DEG C:
=25.0
PLEASE TYPE EXTERNAL AIR TEMPERATURE IN DEG C:
=0.0
RESULTS FOR AIR GAP THICKNESS:0.025
TOTAL HEAT TRANSFER RATE IN W:126.9
HEAT TRANSFER RATE, FRAME, IN W:11.6
HEAT TRANSFER RATE, WINDOW, IN W:115.2
OUTSIDE GLASS TEMPERATURE, IN DEG C: 4.2
INSIDE GLASS TEMPERATURE, IN DEG C:19.5
OUTSIDE FRAME TEMPERATURE, IN DEG C: 1.7
INSIDE FRAME TEMPERATURE, IN DEG C:22.8
   DO YOU WISH TO DO A REPEAT CALCULATION? 1-YES,2-NO:
=2
```

3.2 Effect of a Variable Conductivity in a Plane Slab

In considering the variation of k with temperature in the case of one-dimensional flow in a plane slab, equation (2.3) for the relationship between k and temperature will be used.

For conduction in a single plane slab,

$$q_x = - k_0(1 + \alpha t)\frac{\mathrm{d}t}{\mathrm{d}x}$$

$$\therefore \quad q_x(x_2 - x_1) = -k_0\left[(t_2 - t_1) + \frac{\alpha}{2}(t_2^2 - t_1^2)\right]$$

$$= \frac{- k_0[2(t_2 - t_1) + \alpha(t_2 - t_1)(t_2 + t_1)]}{2}$$

and

$$q_x = \frac{- k_0[2 + \alpha(t_2 + t_1)](t_2 - t_1)}{2(x_2 - x_1)} \tag{3.12}$$

It will be found that equation (3.12) can also be obtained by taking an average of the conductivities at temperatures t_2 and t_1 and substituting into equation (3.3). Equation (3.12) may be used to find the interface temperature between two plane slabs, e.g., for two materials where $k_{1t} = k_{10}(1 + \alpha t)$ and $k_{2t} = k_{20}(1 + \beta t)$. The heat flux through both slabs is the same, hence

$$\frac{-k_{10}[2 + \alpha(t_2 + t_1)](t_2 - t_1)}{2(x_2 - x_1)} = \frac{-k_{20}[2 + \beta(t_3 + t_2)](t_3 - t_2)}{2(x_3 - x_2)}$$

This equation may be solved to find t_2, and then q_x may be calculated.

EXAMPLE 3.3

The heat flux through a plane slab 0·1 m thick is 146 kW/m² for surface temperatures of 120° and 30°C. Find the value and sign of α in the thermal conductivity function given that $k_0 = 0.16$ kW/(mK)

Solution. Using equation (3.12)

$$146 = -0.16\,[2 + \alpha(30 + 120)](30 - 120)/(2 \times 0.1)$$

$$2 + 150\,\alpha = 146 \times 0.2/(0.16 \times 90)$$

$$= 2.03$$

$$\therefore \alpha = +0.03/150 = +2.0 \times 10^{-4}\,K^{-1}$$

3.3 Radial Conduction in Cylindrical Layers

Conduction through thick walled pipes is a common heat transfer problem, and may be treated one-dimensionally if surface temperatures are uniform. The heat flow is then in the radial direction only. Fig. 3.5 illustrates the situation for a single layer. Fourier's law may be applied to a cylindrical layer at radius r:

$$Q_r = -kA\frac{dt}{dr}$$

Here A is the surface area at the radius r, and obviously A will vary between the inner and outer radii. It is therefore convenient to consider a *unit length of cylinder*, when the radial heat transfer is

$$Q_r = - k(2\pi r)\frac{dt}{dr} \qquad (3.13)$$

$2\pi r$ is the area per unit length. Since the same quantity Q_r is flowing through a steadily increasing cylindrical area, the temperature gradient must decrease with increasing radius. Integrating:

$$Q_r \ln \frac{r_2}{r_1} = - 2\pi k(t_2 - t_1)$$

$$\therefore \qquad Q_r = - 2\pi k \frac{(t_2 - t_1)}{\ln r_2/r_1} \qquad (3.14)$$

By analogy with Ohm's law, the thermal resistance per unit length of cylinder in this case is $[\ln (r_2/r_1)]/2\pi k$.

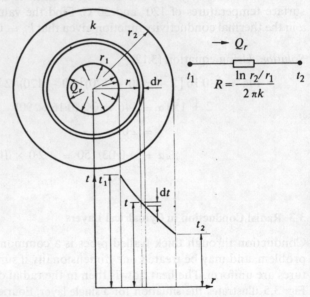

Fig. 3.5. Steady state radial conduction in a cylindrical layer.

Equation (3.14) may also be derived from the general equation for the cylindrical coordinate system (2.16) which simplifies to

$$\frac{d^2t}{dr^2} + \frac{1}{r}\frac{dt}{dr} = 0 \qquad ((2.18))$$

for the case of steady radial conduction in the absence of internal heat generation. This equation may be integrated to give

$$t = C_3 \ln r + C_4 \qquad (3.15)$$

where C_3 and C_4 are constants of integration to be found from boundary conditions. Thus, if $t = t_1$ at $r = r_1$ and $t = t_2$ at $r = r_2$, it is found that

$$t = \frac{(t_2 - t_1)}{\ln r_2/r_1} \ln \frac{r}{r_1} + t_1 \qquad (3.16)$$

To obtain equation (3.14), the temperature gradient is found by differentiating (3.16) and substituting back in (3.13).

A thick walled steam pipe with lagging is a familiar example of multiple cylindrical layers, and the treatment is similar to the multiple plane layer. Fig. 3.6 shows two cylindrical layers separat-

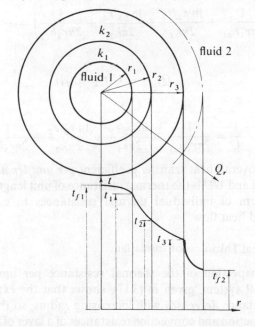

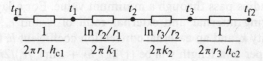

Fig. 3.6. *Steady state radial conduction in concentric cylinders separating two fluids.*

ing two fluids f_1 and f_2. It is assumed that heat transfer at the surfaces is in each case predominantly by convection. Considering unit length, at the inside surface:

$$Q_r = - 2\pi r_1 h_{c1}(t_1 - t_{f1})$$

The same quantity is conducted through the two layers, hence

$$Q_r = - 2\pi k_1 \frac{(t_2 - t_1)}{\ln r_2/r_1} = - 2\pi k_2 \frac{(t_3 - t_2)}{\ln r_3/r_2}$$

It is also convected from the outside surface, so

$$Q_r = - 2\pi r_3 h_{c2}(t_{f2} - t_3)$$

Re-arranging and adding these equations gives:

$$Q_r \left(\frac{1}{2\pi r_1 h_{c1}} + \frac{\ln r_2/r_1}{2\pi k_1} + \frac{\ln r_3/r_2}{2\pi k_2} + \frac{1}{2\pi r_3 h_{c2}} \right) = - (t_{f2} - t_{f1})$$

or

$$Q_r = - U(t_{f2} - t_{f1}) \tag{3.17}$$

where

$$\frac{1}{U} = \frac{1}{2\pi r_1 h_{c1}} + \frac{\ln r_2/r_1}{2\pi k_1} + \frac{\ln r_3/r_2}{2\pi k_2} + \frac{1}{2\pi r_3 h_{c2}}$$

U is the overall heat transfer coefficient *per unit length* between the two fluids and $1/U$ is the thermal resistance of unit length and consists of the sum of individual thermal resistances to conducted and convected heat flow.

3.4 Critical Thickness of Insulation

Closer inspection of the thermal resistance per unit length of a cylindrical system, given in (3.17), shows that the external convection resistance *decreases* with increasing radius, so that the sum of the conduction and convection resistances of a layer of insulation will at some radius pass through a minimum value. For a layer of insulation having internal and external radii of r_i and r_o, a thermal conductivity k, and an external convection coefficient h, the thermal resistance per unit length will be $(1/2\pi r_o h) + ((\ln r_o/r_i)/2\pi k)$ and this will have a minimum value obtained by putting

$$\frac{d}{dr_o}\left(\frac{1}{2\pi r_o h} + \frac{\ln r_o/r_i}{2\pi k}\right) = 0$$

$$\therefore -\frac{1}{r_o^2 h} + \frac{1}{kr_o} = 0$$

or

$$r_o = \frac{k}{h} = \text{critical radius} \qquad (3.18)$$

This is the value of outer radius for which heat transfer through the system will be a maximum. It follows that if for a given k and h the external radius is less than k/h, then increasing the thickness of insulation up to a radius of k/h will *increase*, rather than *decrease* the heat loss from the system. The situation is likely to arise if k has a relatively high value, and h a relatively low value.

3.5 Radial Conduction in Spherical Layers

Another simple instance of one-dimensional conduction is that which can occur in a spherical layer. Conduction will be only in the radial direction if the temperatures of the two spherical surfaces are uniform. The radial conduction is given by:

$$Q_r = -4\pi k \frac{r_1 r_2}{r_2 - r_1} \cdot (t_2 - t_1) \qquad (3.19)$$

and the overall heat transfer coefficient for a double spherical layer separating two fluids f_1 and f_2 is

$$1\bigg/\left(\frac{1}{4\pi r_1^2 h_{c1}} + \frac{1}{4\pi k_1} \cdot \frac{r_2 - r_1}{r_1 r_2} + \frac{1}{4\pi k_2} \cdot \frac{r_3 - r_2}{r_2 r_3} + \frac{1}{4\pi r_3^2 h_{c2}}\right)$$

The value of the critical r_0 for a sphere is $2k/h$.

EXAMPLE 3.4

In this program the critical radius of an insulated pipe may be investigated. The internal radius of the insulation R1, and the internal and external temperatures T1 and TA may be regarded as fixed in any one problem, and the effects of the insulation thickness TH, insulation thermal conductivity TK, and the outer convection coefficient H, may be investigated.

The BASIC program list and sample results are given below. In the example $R1 = 0.1$ m, $TK = 0.85$ W/m K, and $H = 5.0$ W/m^2 K. This gives a critical outer radius of 0.17 m and, for insulation thicknesses between 25 and 200 mm, the heat transfer variation is given in the following table:

Insulation thickness, mm	Outer radius, m	Heat transfer, W/m
25	0·125	0·607
40	0·140	0·620
55	0·155	0·626
70	0·170	0·628
90	0·190	0·625
200	0·300	0·577

BASIC Program Listing

```
10      R1=0.1
20      T1=200.0
30      TA=20.0
40      PRINT,"PLEASE TYPE INSULATION THICKNESS, MM:"
50      INPUT TH
60      R2=TH/1000.0+R1
70      PRINT,"PLEASE TYPE INSULATION THERMAL CONDUCTIVITY, W/M DEG C:"
80      INPUT TK
90      PRINT,"PLEASE TYPE OUTER RADIUS CONVECTION COEFFICIENT, W/M2 DEG C:"
100     INPUT H
110     RCRIT=TK/H
120     RESIS=1.0/(6.2832*R2*H)+ALOG(R2/R1)/(6.2832*TK)
130     QFLOW=(T1-TA)/RESIS
140     PRINT,"RESULTS FOR AN OUTER RADIUS OF, M:",R2
150     PRINT,"CRITICAL RADIUS FOR THE COEFFICIENTS USED:",RCRIT
160     PRINT,"HEAT FLOW, W/M PIPE LENGTH:",QFLOW
170     PRINT,"DO YOU WISH TO DO A REPEAT RUN? 1-YES; 2-NO"
180     INPUT IX%
190     IF(IX%<2)GO TO 40
200     STOP
```

Sample Result from Listing

```
PLEASE TYPE INSULATION THICKNESS, MM:
=25.0
PLEASE TYPE INSULATION THERMAL CONDUCTIVITY, W/M DEG C:
=0.85
PLEASE TYPE OUTER RADIUS CONVECTION COEFFICIENT, W/M2 DEG C:
=5.0
RESULTS FOR AN OUTER RADIUS OF, M:  0.12500000E 00
CRITICAL RADIUS FOR THE COEFFICIENTS USED:  0.17000000E 00
HEAT FLOW, W/M PIPE LENGTH:  0.60722832E 03
DO YOU WISH TO DO A REPEAT RUN? 1-YES; 2-NO
=1
PLEASE TYPE INSULATION THICKNESS, MM:
=70.0
```

```
PLEASE TYPE INSULATION THERMAL CONDUCTIVITY, W/M DEG C:
=0.85
PLEASE TYPE OUTER RADIUS CONVECTION COEFFICIENT, W/M2 DEG C:
=5.0
RESULTS FOR AN OUTER RADIUS OF, M:  0.17000000E 00
CRITICAL RADIUS FOR THE COEFFICIENTS USED:  0.17000000E 00
HEAT FLOW, W/M PIPE LENGTH:  0.62806211E 03
DO YOU WISH TO DO A REPEAT RUN? 1-YES; 2-NO
=1
PLEASE TYPE INSULATION THICKNESS, MM:
=200.0
PLEASE TYPE INSULATION THERMAL CONDUCTIVITY, W/M DEG C:
=0.85
PLEASE TYPE OUTER RADIUS CONVECTION COEFFICIENT, W/M2 DEG C:
=5.0
RESULTS FOR AN OUTER RADIUS OF, M:  0.30000000E 00
CRITICAL RADIUS FOR THE COEFFICIENTS USED:  0.17000000E 00
HEAT FLOW, W/M PIPE LENGTH:  0.57727842E 03
DO YOU WISH TO DO A REPEAT RUN? 1-YES; 2-NO
=2
```

3.6 Conduction with Heat Sources

The flow of electricity in a material gives rise to ohmic heating and, generally, the resulting heat flow is at least two-dimensional. However, if the flow of current in a flat wide bar, or the heating of a flat plate by eddy currents is being considered, then the heat flow is essentially one-dimensional if edge effects are neglected. (See Fig. 3.7. where $1 \gg b$).

The general equation for the rectangular coordinate system, when applied to this problem, reduces to

$$0 = \alpha \left(\frac{d^2 t}{dx^2} \right) + \frac{q'}{\rho c_p} \qquad ((2.10))$$

Assuming q' uniform in space, equation (2.10) is integrated to give

$$t = - \frac{q' x^2}{2k} + C_5 x + C_6 \qquad (3.20)$$

where C_5 and C_6 are constants of integration to be determined from boundary conditions.

If the boundary condition is convection to a known fluid temperature, so that the solid boundary temperatures, t, are unknown, two further equations are obtained by equating conduction to convection at the boundary, for example

$$h_a (t_{x_1} - t_a) = - \left(-k \left(\frac{dt}{dx} \right)_{x_1} \right) \qquad (3.21)$$

and

$$h_b\,(t_{x_2} - t_a) = -\,k\left(\frac{dt}{dx}\right)_{x_2} \qquad (3.22)$$

where h_a and t_a apply at face x_1 and h_b and t_b apply at face x_2. Four equations are now available to give C_5, C_6 and t_{x_1} and t_{x_2}.

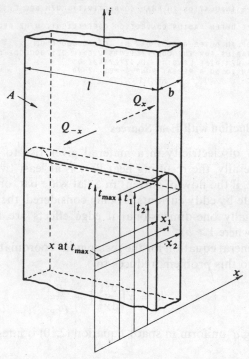

Fig. 3.7.   One-dimensional conduction in a plane layer with internal heat generation.

The heat transfer at any plane, x, is obtained by differentiating equation (3.20) and applying Fourier's law. Thus

$$q_x = -\,k\left(\frac{dt}{dx}\right)_x \qquad (3.23)$$

If $dt/dx = 0$, the temperature is a maximum and the heat flux is zero. Thus if one face of the slab is insulated, it will also be the hottest.

The maximum temperature is found by putting the value of x at which $dt/dx = 0$ into equation (3.20).

One-dimensional conduction in the radial direction will occur in a rod or hollow cylindrical bar if surface temperatures are uniform. The maximum temperature will occur at the centre of a rod, and at an intermediate radius in a hollow bar if both surfaces are cooled.

The general equation in cylindrical coordinates (2.16) reduces to

$$\alpha\left(\frac{d^2t}{dr^2} + \frac{1}{r}\frac{dt}{dr}\right) + \frac{q'}{\rho c_p} = 0 \qquad ((2.17))$$

for this situation. The solution of this is

$$t = -\frac{q'r^2}{4k} + C_7 \ln r + C_8 \qquad (3.24)$$

which may be obtained by making the substitution $dt/dr = p$. Values of C_7 and C_8, the constants of integration, may be found by substituting the known boundary conditions, see Fig. 3.8. The value

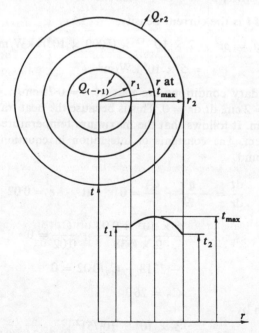

Fig. 3.8. *Radial conduction in a cylindrical layer with internal heat generation.*

of r at which $dt/dr = 0$ gives the position of the maximum temperature, and this substituted in (3.24) gives the value of the maximum temperature.

The following example illustrates the way in which ohmic heating problems may be solved.

EXAMPLE 3.5

An internally cooled copper conductor of 4 cm outer diameter and 1·5 cm inner diameter carries a current density of 5000 amp/cm². The temperature of the inner surface is maintained at 70°C, and it may be assumed that no heat transfer takes place through insulation surrounding the copper. Determine the equation for temperature distribution through the copper, hence find the maximum temperature of the copper, the radius at which it occurs, and the heat transfer rate internally. Check that this is equal to the total energy generation in the conductor. For copper, take $k = 0.38 \, \text{kW/(mK)}$ and the resistivity $\rho = 2 \times 10^{-11}$ ohm metre.

Solution. If i is the current density,

$$q' = \rho i^2 = 2 \times 10^{-11} \times (5000 \times 10^4)^2 \, \text{kW/m}^3$$

$$= 5 \times 10^4 \, \text{kW/m}^3$$

The boundary conditions are that at $r = 0.75$ cm, $t = 70°C$ and that at $r = 2$ cm, $dt/dr = 0$. This is because the heat transfer is zero at $r = 2$ cm. It follows that the maximum temperature also occurs at $r = 2$ cm. The constants of integration in equation (3.24) may now be found.

$$\frac{dt}{dr} = -\frac{q'r}{2k} + \frac{C_7}{r} = 0 \quad \text{at} \quad r = 0.02$$

$$\therefore \quad -\frac{5 \times 10^4 \times 0.02}{2 \times 0.38} + \frac{C_7}{0.02} = 0$$

$$\therefore \quad -1318 + C_7/0.02 = 0$$

$$C_7 = 26.3$$

C_8 is given by

$$70 = -\frac{5 \times 10^4}{4 \times 0.38} \times \left(\frac{0.75}{100}\right)^2 + 26.3 \ln(0.0075) + C_8$$

$$= -1 \cdot 85 - 128 \cdot 7 + C_8$$
$$\therefore \qquad C_8 = 200 \cdot 6$$

The equation for temperature is therefore:

$$t = -32{,}900 r^2 + 26 \cdot 3 \ln r + 200 \cdot 6$$

with r in metres.

The maximum temperature occurs at the outer radius. Substituting $r = 0 \cdot 02$ m in the above equation gives

$$t_{max} = -13 \cdot 17 - 102 \cdot 8 + 200 \cdot 6$$
$$= 84 \cdot 6°C$$

To calculate the heat transfer rate internally, it is first necessary to find the temperature gradient at $r = 0 \cdot 0075$ m. Thus

$$\left(\frac{dt}{dr}\right)_{r=0 \cdot 0075} = -\frac{5 \times 10^4 \times 0 \cdot 0075}{2 \times 0 \cdot 38} + \frac{26 \cdot 3}{0 \cdot 0075}$$
$$= -494 + 3510$$
$$= +3016$$

The heat transfer internally is in the direction of negative radius, hence

$$Q_{(-r)} = -\left(-kA\frac{dt}{dr}\right)$$
$$= +0 \cdot 38 \times (2\pi \times 0 \cdot 0075) \times 3016$$
$$= 53 \cdot 9 \text{ kW/m length}$$

This result may be checked since all the heat generated in the conductor must be dissipated internally.

$$\therefore \qquad Q_{(-r)} = (\text{volume/m length}) \times q'$$
$$= \pi(0 \cdot 02^2 - 0 \cdot 0075^2) \times 5 \times 10^4$$
$$= 53 \cdot 9 \text{ kW/m length.}$$

PROBLEMS

1. The walls of a refrigerator for a shop consist of slag wool $0 \cdot 1522$ m thick sandwiched between sheet iron, $0 \cdot 0794$ cm thick, on one side and asbestos

board, 0·953 cm thick, on the other. The total surface effective for heat transfer is 37·2 m². The atmospheric temperature is 18·3°C and the temperature in the cold room is −3·9°C.

The thermal conductivity of iron, slag wool, and asbestos board may be taken as 69·1, 0·346, 1·21 × 10⁻³ respectively and the surface heat transfer coefficient as 1·705 × 10⁻³; in kW, m, K, units.

Compute the heat leakage into the refrigerator. (Ans. 0·51 kW.) (*King's College, London*).

2. A spherical container 1·22 m internal diameter is made of sheet metal of negligible thermal resistance and covered by cork insulation 0·457 m thick. The interior contains a liquefied gas at −62·2°C for which a surface heat transfer coefficient of 1·06 kW/(m² K) may be considered to apply. The atmospheric temperature is 18·3°C. Moisture vapour permeates the cork and freezes at a suitable position to form an ice barrier. The mean surface co-efficient for the outside may be regarded as 0·021 kW/(m² K). Calculate the thickness of the ice assuming that the conduction characteristics of the cork remain constant throughout.

Assume the thermal conductivity for cork is 43·2 × 10⁻⁶ kW/(m K). (Ans. 0·305 m.) (*Queen Mary College, London*).

3. A 30 mm diameter pipe at 100°C is losing heat by natural convection to the atmosphere at 20°C at the rate of 0·1 kW per m length. It is required to cut down this loss to 0·05 kW/m. Two insulating materials A and B are available. There is sufficient of A to use it at the rate of 3·14 × 10⁻³ m³/m length, and of B to use it at the rate of 4·0 × 10⁻³ m³/m. The thermal conductivities of A and B are 0·005 and 0·001 kW/(m K) respectively. Is it possible to achieve the required degree of insulation? Assume the convection coefficient applicable to the bare pipe is also applicable to the outer surface of insulation. (Ans. B inside 0·0437 kW/m; A inside 0·0742 kW/m.) (*The City University*)

4. Calculate the surface temperature and the maximum temperature of a 10 mm diameter steel conductor carrying 5000 amps and forced convection cooled to the atmosphere at 15°C with a convection coefficient of 5·55 kW/(m² K). For the conductor, take the electrical resistivity as 8 × 10⁻⁸ ohm m, and the thermal conductivity as 0·12 kW/(m K). (Ans. 161·3°C and 178·2°C.)

5. (i) Define the term thermal resistance and show that, when heat flows through a number of individual resistances in series, the overall resistance is equal to the sum of the individual resistances.
(ii) A double-glazed window consists of two sheets of glass separated by a gap. The gap is filled with a gas, but is sufficiently thin to prevent convection between the two sheets of glass. The area of the window in elevation is A, the thickness of each sheet of glass is x and the thickness of the gap is y. The thermal conductivities of the glass and of the gas in the gap are k_x and k_y respectively. The surface heat-transfer coefficients inside and outside the building are h_1 and h_2 respectively; the corresponding air temperatures are t_1 and t_2. Neglecting radiation, obtain an expression for the heat transfer rate q, in terms of A, x, y, k_x, k_y, h_1, h_2, t_1, and t_2.

(iii) Find the percentage reduction in heat loss when a single-glazed window is replaced by a double-glazed window. Assume that the values of A, x, k_x, h_1, h_2, t_1, and t_2 are the same for both windows, the symbols having the same meaning as in section (ii). Numerical data:

$x = 0.318$ cm; $y = 0.635$ cm;

$k_x = 865 \times 10^{-6}$ kW/(m K) $k_y = 26 \times 10^{-6}$ kW/(m K)

$h_1 = 8.52 \times 10^{-3}$ kW/(m^2 K) $h_2 = 14.2 \times 10^{-3}$ kW/(m^2 K)

(Ans. 56·5 per cent)   (*Imperial College, London*).

6. The inner surface of a 0·23 m furnace wall is at 800°C. The outer surface convects to the atmosphere at 21°C, with a coefficient of 0·012 kW/(m^2 K). The conductivity of the furnace wall is 870×10^{-6} kW/(m K). To cut down heat loss, an additional wall 0·23 m thick of insulating brick is added on the outside, having a conductivity of 260×10^{-6} kW/(m K). For the same outer surface coefficient, calculate the percentage reduction of heat loss, the brick interface temperature, and the brick outer surface temperature. (Ans. 71·1 %, 633°C, 73°C.)

7. A wide copper strip 10 mm thick carries an electric current of density 50 amp/mm^2. The heat generated is dissipated from the two wide faces of the strip by convection. On the left-hand face the convection coefficient is 5 kW/m^2 K and on the right it is 10 kW/m^2 K. The surroundings are at 25°C.

You may assume for one-dimensional conduction with heat generation, that

$$\frac{d^2 t}{dx^2} = - \frac{q_g}{k}$$

where the symbols have their usual meanings.
Calculate:
(i)   The constants of integration in the temperature distribution equation for the copper strip.
(ii) The temperatures of the two faces of the copper strip.
(iii) The heat fluxes (kW/m^2) at the two faces, and show the sum of the fluxes equals the heat generation in the volume of copper per unit area of wide face.
For copper, $\rho = 2 \times 10^{-11}$ ohm metre, $k = 0.38$ kW/(m K).
(Ans. (i) 456·32 K/m; (ii) 59·67°C and 57·66°C. (iii) 173·4 and 326·6 kW/m^2.)

8. The walls of a house consist of two skins of brick each 115 mm thick, separated by an air gap of 50 mm. The heat transfer coefficients on the outside and inside walls of the house, respectively, are:

       outside   inside

by convection 7 W/m^2 K;   by convection 3 W/m^2 K
by radiation   5 W/m^2 K;   and   by radiation   4 W/m^2 K

Across the air gap, the overall heat transfer coefficient is 4 W/m² K, taking into account conduction, radiation and convection effects.

To reduce heat loss the air gap is filled with insulation, so that heat transfer across this region is now by conduction only.

(a) For a wall area of 70 m² calculate, for the uninsulated wall, the annual heat loss in kWh, assuming an annual mean ambient temperature of 12°C and a mean internal temperature of 22°C maintained 24 hours a day.

(b) Calculate the percentage reduction in annual heat loss across the walls by the insertion of insulation, of thermal conductivity 0·032 W/mK.

The thermal conductivity of brick is 0·81 W/m K.

(Ans. (a) 8066·3 kWh. (b) 63·32 per cent.)

9. A hot water storage cylinder consists of a copper cylinder with hemispherical ends. The cylindrical section is 0·4 m diameter by 0·6 m long, and the hemispherical ends have a radius of 0·2 m, so that the overall length is 1·0 m. The copper surface has a temperature of 60°C, and there is a surface heat transfer coefficient of 5·4 W/m² K to the ambient and surroundings at 18°C.

To reduce heat loss, 100 mm of insulation is added to the whole surface, for which the thermal conductivity is 0·04 W/m K. The surface coefficient of 5·4 W/m² K continues to act now on the outside surface of the insulation.

(i) Calculate the original heat loss rate for the whole tank and the percentage reduction in heat loss by adding the insulation.

(ii) Calculate the temperature on the outside surface of the insulation, both on the cylindrical surface and the hemispherical surface. Why are the two values not the same?

You may assume that the thermal resistance of a spherical layer of radii r_1 and r_2 and of thermal conductivity k is given by $(r_2 - r_1)/(4\pi k r_1 r_2)$.

(Ans. (i) 285 W; 90·6 per cent reduction. (ii) Sides 20·41°C; ends 19·98°C.)

4

Two-dimensional steady state conduction

It is important to realize that in many cases a conduction problem is over simplified by the use of one-dimensional treatment, which means the neglect of edge and corner effects which must be present in any finite object. The error involved in this neglect will depend on the dimensions of the system. Consider, for example, the wall of a building some 6 m long and 200 mm thick. In the absence of doors and windows, conduction through such a wall will be one-dimensional over the greater part of the 6 m length and the error involved in neglecting the corner effects will not be great. In contrast, conduction through a chimney, say, 300 mm square internally and 1 m square externally, is *essentially* two-dimensional. Again a simplifying assumption is being made, since near the base and top of the chimney conduction will be three-dimensional. Thus those problems will be considered in this chapter which may be assumed to be two-dimensional without introducing significant error. This will cover the majority of heat conduction problems which are sufficiently simple to include in an introductory text.

Two-dimensional problems in rectangular coordinates only are to be considered. The two equations, with and without heat generation, are:

$$\alpha\left(\frac{\partial^2 t}{\partial x^2} + \frac{\partial^2 t}{\partial y^2}\right) + \frac{q'}{\rho c_p} = 0 \qquad ((2.9))$$

$$\frac{\partial^2 t}{\partial x^2} + \frac{\partial^2 t}{\partial y^2} = 0 \qquad ((2.11))$$

Solutions to these equations are, of course, possible, but the more readily obtained ones depend on the choice of somewhat unrepresentative configurations or boundary conditions. As an alternative,

39

therefore, numerical procedures will be described. These have an advantage of being applicable to any two-dimensional shape. The first method, involving simple arithmetic, is suited to problems having specified boundary temperatures. The second method, which can consider more general and complicated boundary conditions, involves the use of a digital computer.

The main difference between the analytical solution and numerical methods is that the former will give an equation from which the temperature may be obtained anywhere in the solid, whereas the latter will give values of temperature at chosen specific points only. The accuracy will depend on how close together are the chosen points; however, many points will entail much more work than a few.

4.1 A Numerical Solution of Two-dimensional Conduction

A numerical method involving a process known as *relaxation*[1] will be introduced by consideration of a typical example of two-dimensional conduction, the right-angled corner. The method is suited to simple problems involving only a few specific points. For conduction fields involving many points, some elementary computing procedures are considered in Section 4.2.

A right-angled corner, forming part of a square hollow section, such as a chimney, is shown in Fig. 4.1. For boundary temperatures uniform on the inside and outside surfaces, a one-eighth unique part exists as shown, involving only 11 mesh points in this example.

It is necessary to replace the differential equation (2.11) by finite difference approximations relating temperatures around a mesh point, and this is possible if temperatures vary continuously in the x- and y- directions, expressible as $t = f(x)$ and $t = f(y)$. Using MacLaurin's series, the temperatures at points 1 and 3 may be expressed in terms of t_0 at point 0, the differential coefficients of $t = f(x)$ at $x = 0$ at point 0, and the mesh size a. Thus:

$$\text{at } x = +a, t_1 = t_0 + \left(\frac{\partial t}{\partial x}\right)_0 \frac{a}{1!} + \left(\frac{\partial^2 t}{\partial x^2}\right)_0 \frac{a^2}{2!} + \left(\frac{\partial^3 t}{\partial x^3}\right)_0 \frac{a^3}{3!} + \cdots$$

$$(4.1)$$

$$\text{at } x = -a, t_3 = t_0 - \left(\frac{\partial t}{\partial x}\right)_0 \frac{a}{1!} + \left(\frac{\partial^2 t}{\partial x^2}\right) \frac{a^2}{2!} - \left(\frac{\partial^3 t}{\partial x^3}\right) \frac{a^3}{3!} + \cdots$$

$$(4.2)$$

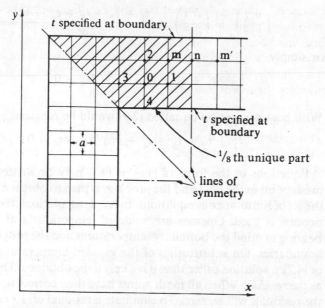

Fig. 4.1. Construction for a numerical solution of two-dimensional conduction, in a quarter of a hollow square section.

neglecting higher powers. When added together these give

$$t_1 + t_3 = 2t_0 + \left(\frac{\partial^2 t}{\partial x^2}\right)_0 a^2$$

or

$$\left(\frac{\partial^2 t}{\partial x^2}\right)_0 = \frac{t_1 + t_3 - 2t_0}{a^2} \tag{4.3}$$

Writing similar equations for t_2 at $y = +a$ and t_4 at $y = -a$, it is possible to obtain in like manner

$$\left(\frac{\partial^2 t}{\partial y^2}\right)_0 = \frac{t_2 + t_4 - 2t_0}{a^2} \tag{4.4}$$

Both (4.3) and (4.4) contain *discretization errors* involving terms containing fourth and higher powers of a. Equation (2.11) may thus be replaced by

$$\left(\frac{\partial^2 t}{\partial x^2}\right)_0 + \left(\frac{\partial^2 t}{\partial y^2}\right)_0 = \frac{t_1 + t_2 + t_3 + t_4 - 4t_0}{a^2} = 0 \qquad (4.5)$$

or simply

$$t_1 + t_2 + t_3 + t_4 - 4t_0 = 0 \qquad (4.6)$$

With heat generation equation (2.9) would be replaced by

$$t_1 + t_2 + t_3 + t_4 + a^2 q'/k - 4t_0 = 0 \qquad (4.7)$$

Equations of the form of (4.6) or (4.7) may be written for every mesh point in the field and the problem is then to obtain a solution of the set of simultaneous equations. In hand calculations, the *relaxation* process is used. Guesses are made of temperatures at the points, bearing in mind the boundary temperatures and the proximity to the boundaries. On substitution of the guessed temperatures into (4.6) or (4.7), a solution other than 0 is likely to be obtained. This is known as the *residual*. When all mesh points have their correct temperatures, the residuals will be zero. To eliminate a residual of $+r$ or $-r$, it will be seen that t_0 must be increased by $+r/4$ or $-r/4$ respectively. Carrying out this operation will alter the residuals at surrounding field points by the same amount, i.e., $+r/4$ or $-r/4$. Care must be exercised around lines of symmetry. Thus $t_m = t_{m'}$ in Fig. 4.1, and eliminating a residual of $\pm r$ at m will alter the residual at n by $\pm r/2$ and not by $\pm r/4$. Eliminating a residual at n will have the normal effect on the residual at m and m'.

To carry out the relaxation process, the initial residuals due to guessed temperatures are calculated, then the residuals are eliminated one by one starting with the largest. Subsequent operations may re-introduce residuals previously eliminated, so that some points must be treated more than once. The process generally continues until residuals are ± 2 or smaller, indicating temperatures are within $\frac{1}{2}°$ of their correct value. The magnitude of error arising depends on the overall temperature range involved.

To illustrate a step in the procedure, let $t_4 = 120°$ (boundary value), $t_0 = 80°$, $t_2 = 40°$, $t_1 = 100°$ and $t_3 = 70°$ as original guesses in Fig. 4.1. Then equation (4.6) gives: $100 + 40 + 70 + 120 - 320 = +10$. The residual of $+10$ could be reduced to -2 by adding $3°$ onto t_0 to give $83°$. Residuals at points 1, 2 and 3 are then increased by $+3$. If heat generation is present the constant term $a^2 q'/k$ is included in the equation. It does not enter into the relaxation procedure, but it is seen to have the effect of raising the general level of temperature.

Once the temperatures are known, heat conduction in the section is obtained from a summation of conduction along the mesh lines. Heat transfer through unit length of the whole section would be the average of heat conduction in and out at the hotter and cooler boundaries respectively. Now imagine point 0 to be at the *centre* of a square of size a at a temperature t_0. Similarly there are squares at t_1, t_2 and t_3, etc., and at the boundary there is a half-square at t_4. Heat conduction between 4 and 0 per unit length of section is given by

$$Q_{(4-0)} = - k(a \times 1) \frac{(t_0 - t_4)}{a} = k(t_4 - t_0) \, \text{kW/m} \qquad (4.8)$$

Total heat conduction at the boundary is a summation of similar terms.

When there is heat generation in the section, heat transfer out of the section at point 4 is given by:

$$Q_{(0-4)} + q'a^2/2 = k(t_0 - t_4) + q'a^2/2 \qquad (4.9)$$

Similarly heat transfer into the section at point 4 is:

$$Q_{(4-0)} - q'a^2/2 = k(t_4 - t_0) - q'a^2/2 \qquad (4.10)$$

These relationships arise from an energy balance on the boundary half square, i.e., heat generated = net heat flow out.

EXAMPLE 4.1

Establish the temperatures a–k in the duct shown, by relaxation, and calculate the conduction heat transfer through the duct. k for the duct is $0 \cdot 1 \, \text{kW/(mK)}$.

Solution. Take initial guesses for the temperatures as: a = 50°, b = 48°, c = 41°, d = 25°, e = 41°, f = 46°, g = 47°, h = 48°, j = 50°, k = 50°. Residuals are calculated first, e.g., for point a: 80 + 20 + 48 + 48 − (4 × 50) = −4. The relaxation table with initial residuals shown is as follows:

Mesh point:	a	b	c	d	e	f	g	h	j	k
Initial residual:	−4	−1	+9	+22	+7	+4	+6	+5	−2	0

Operation:										
+6 at d = 31°				+15	−2	+13				
+4 at c = 45°			+3	−1	+2					
+3 at e = 44°					+5	+1	+7			
+2 at f = 48°						+3	−1	+8		
+2 at g = 49°							+1	0	+7	
+2 at h = 50°								+2	−1	0
+1 at d = 32°				0	+1	+4				
+1 at e = 45°					+2	0	+2			
−1 at a = 49°	0	+2								
+1 at f = 49°						+1	−2	+3		
+1 at g = 50°	0	+2	0	+2	+1	−1	−1	0	0	0

Final temperatures are: a = 49°, b = 48°, c = 45°, d = 32°, e = 45°, f = 49°, g = 50°, h = 50°, j = 50°, k = 50°.

For the whole duct, heat conduction in = $4[\frac{1}{2}k(80-49) + k(80-48) + k(80-45) + k(80-45) + k(80-49) + k(80-50) + k(80-50) + k(80-50) + \frac{1}{2}k(80-50)] = 101.4\ kW/m.$

Heat conduction out = $4[\frac{1}{2}k(49-20) + k(48-20) + k(45-20) + 2k(32-20) + k(45-20) + k(49-20) + k(50-20) + k(50-20) + k(50-20) + \frac{1}{2}k(50-20)] = 100.2\ kW/m.$ The average figure is 100.8 kW/m.

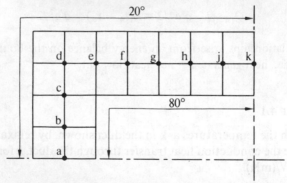

Fig. Example 4.1. Unique part of rectangular duct.

4.2 Iterative Computing Procedures for Two-dimensional Steady State Conduction

The relaxation procedure is suitable for fields of few mesh points, and is most easily performed by hand. In an iterative process, it is the new value of a temperature calculated at a mesh point, in comparison with the old value, that is of interest, rather than the residual obtained at the point when the temperature equation is solved. For small fields, individual programs may be written for each problem, as in Example 4.2.

EXAMPLE 4.2

The figure shows a hollow square section, with the centre hollow rotated through 45°. A square mesh is superimposed so that temperatures in the section may be determined. Then by considering heat flow along mesh lines terminating at the boundaries, the heat flow into and out of the section is found, per unit length. A simple program in BASIC is given to carry out this task. The principles involved may be deduced by studying the listing. All points are solved to form one complete iteration, and iterations continue until all points have changed in temperature by less than 0·1°C since the previous iteration. In this example this is achieved in 9 iterations. The average heat flow for a temperature drop of 80°C is 1510·54 W/m, so that the actual surface values vary by 6·67 per cent. This may be attributed to the coarse mesh with the inside surface set at 45° to the mesh lines.

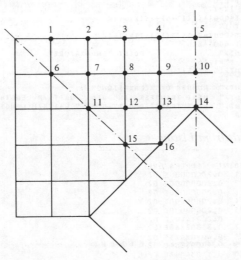

Fig. Example 4.2.

BASIC Program Listing

```
10      DIMENSION T(16),TT(16),DT(16)
20      TK=2.32
30      FOR I%=1 TO 16:READ T(I%):NEXT I%
40      DATA 20.0,20.0,20.0,20.0,20.0,50.0,50.0,50.0,50.0,50.0,50.0,50.0,
```

```
50      &50.0,100.0,50.0,100.0
60      NX=0
70      NX=NX+1
80      TT(6)=2.0*(T(1)+T(7))/4.0
90      DT(1)=ABS(TT(6)-T(6))
100     T(6)=TT(6)
110     TT(10)=(T(5)+T(14)+2.0*T(9))/4.0
120     DT(2)=ABS(TT(10)-T(10))
130     T(10)=TT(10)
140     TT(11)=2.0*(T(7)+T(12))/4.0
150     DT(3)=ABS(TT(11)-T(11))
160     T(11)=TT(11)
170     TT(15)=2.0*(T(12)+T(16))/4.0
180     DT(4)=ABS(TT(15)-T(15))
190     T(15)=TT(15)
200     FOR IX=7,9
210     TT(IX)=(T(IX-1)+T(IX+1)+T(IX-5)+T(IX+4))/4.0
220     DT(IX-2)=ABS(TT(IX)-T(IX))
230     T(IX)=TT(IX)
235     NEXT IX
240     FOR IX=12,13
250     TT(IX)=(T(IX-1)+T(IX+1)+T(IX-4)+T(IX+3))/4.0
260     DT(IX-4)=ABS(TT(IX)-T(IX))
270     T(IX)=TT(IX)
275     NEXT IX
280     LX=0
290     FOR IX=1,9
300     IF(DT(IX).LT.0.1)GO TO 320
310     LX=LX+1
320     NEXT IX
330     IF(LX.GT.0)GO TO 70
340     QIN=0.5*(T(14)-T(10)+T(14)-T(13))*TK
350     QIN=QIN+(T(16)-T(15)+T(16)-T(13))*TK
360     QIN=8.0*QIN
370     QOUT=0.0
380     FOR IX=1,4
390     QOUT=QOUT+(T(IX+5)-T(IX))*TK
400     NEXT IX
410     QOUT=QOUT+0.5*(T(10)-T(5))*TK
420     QOUT=QOUT*8.0
430     PRINT,"              POINT TEMPERATURE"
435     FOR IX=1,16
440     PRINT,IX,T(IX)
445     NEXT IX
450     PRINT," NUMBER OF RELAXATIONS",N
460     PRINT," HEAT TRANSFER IN",QIN," WATTS/M LENGTH"
465     PRINT," HEAT TRANSFER OUT",QOUT," WATTS/M LENGTH"
470     STOP
```

Sample Result from Listing

```
POINT  TEMPERATURE
  1    0.20000000E 02
  2    0.20000000E 02
  3    0.20000000E 02
  4    0.20000000E 02
  5    0.20000000E 02
  6    0.26914492E 02
  7    0.33382788E 02
  8    0.40908473E 02
  9    0.43095284E 02
 10    0.54034888E 02
 11    0.47755155E 02
 12    0.61735517E 02
 13    0.77457700E 02
 14    0.10000000E 03
 15    0.80840663E 02
 16    0.10000000E 03
NUMBER OF RELAXATIONS
HEAT TRANSFER IN  0.14097312E 04 WATTS/M LENGTH
HEAT TRANSFER OUT 0.16113510E 04 WATTS/M LENGTH
```

The simplest possible case has been considered so far, i.e. that of fixed boundary temperatures. If these are uniform then the boundary is said to be *isothermal*. What happens beyond the boundary to create the isothermal condition is outside the scope of the problem, and in this sense the exercise is rather unrealistic. A boundary which is convecting or radiating, or perhaps is insulated, is the more practical situation. Mesh points occurring on such boundaries will have temperature relationships other than (4.6) or (4.7) for points in the field. Also, internal boundaries between different conducting materials may exist. With these and other complexities such as the boundary shape, the work soon becomes too complicated to be treated by relaxation methods. However, whatever these complexities may be, the problem always reduces to solving a set of simultaneous equations, and two computer-based methods are available. The first is that of a direct solution using the Gaussian elimination method[3], and the second is an iterative solution. The essentials of the second method will be described since one basic program with minor changes may be applied to a wide range of problems.

4.2.1 Mesh Point Temperature Relationships for Boundary Points

The first step in preparing a computer program is to consider all the different mesh point temperature relationships that will be involved. Later these can be translated into a form suitable for the program. The relationships to be used are based on the electrical resistance analogue, a technique used extensively for conduction problems prior to the availability of computers. It should be pointed out that in some instances this method produces the same result as the true finite difference relationship (e.g. equations (4.6) and (4.11) below), but when a difference does occur the order of accuracy is lower. For further discussion the reader is referred to Bayley, Owen, and Turner[2].

To introduce the method involved, consider the field point shown in Fig. 4.2. The square mesh is of size a, and the material has thermal conductivity k. The resistances between the centres of squares 1 to 4 and square 0 are therefore equal, and are $a/(k \times a \times 1)$ for unit thickness of the field. The conduction heat transfer (or 'current') across this resistance, for temperature t_1 at point 1, and t_0 at point 0 is therefore $[-(k \times a \times 1)/a](t_0 - t_1) = k(t_1 - t_0)$. The summation of heat transfers from all mesh points to point 0 must be zero in steady state and hence

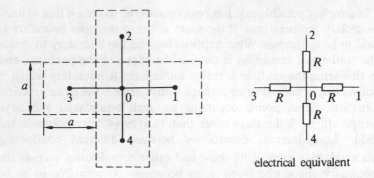

electrical equivalent

Fig. 4.2. Field node in two-dimensional steady state conduction.

$$k(t_1 - t_0) + k(t_2 - t_0) + k(t_3 - t_0) + k(t_4 - t_0) = 0$$
$$t_1 + t_2 + t_3 + t_4 - 4t_0 = 0 \qquad ((4.6))$$

Thus equation (4.6) has been confirmed by this method.

Some representative examples of boundary mesh points will now be given.

Convecting Boundary. Fig. 4.3 shows the physical situation at a convecting boundary. **Double resistances (or half conductances)**

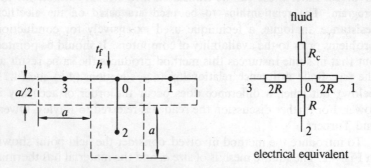

electrical equivalent

Fig. 4.3. Boundary node with convection.

exist between points 1 and 3 and point 0, and a convection resistance R_c exists beyond the solid boundary, of magnitude $1/h(a \times 1)$. The energy balance is

$$\frac{k(t_1 - t_0)}{2} + \frac{k(t_3 - t_0)}{2} + k(t_2 - t_0) + ha(t_f - t_0) = 0$$

$$\therefore \frac{t_1 + t_3}{2} + t_2 + (ha/k)t_f - (2 + ha/k)t_0 = 0 \qquad (4.11)$$

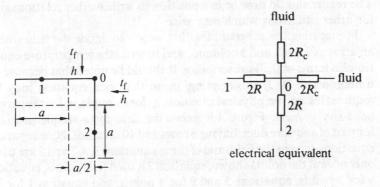

Fig. 4.4.   *External boundary corner with convection.*

Boundary Corners with Convection. An external corner is shown in Fig. 4.4. and the energy balance is given by

$$\frac{k(t_1 - t_0)}{2} + \frac{k(t_2 - t_0)}{2} + \frac{ha}{2}(t_f - t_0) + \frac{ha}{2}(t_f - t_0) = 0$$

$$\therefore t_1 + t_2 + (2ha/k)\, t_f - (2 + 2ha/k)\, t_f = 0 \qquad (4.12)$$

For the internal corner shown in Fig. 4.5, it is left to the reader to show that

$$[(t_1 + t_2)/2] + t_3 + t_4 + (ha/k)\, t_f - (3 + ha/k)\, t_0 = 0 \ (4.13)$$

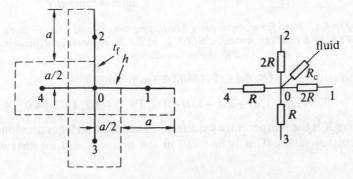

Fig. 4.5.   *Internal boundary corner with convection.*

4.2.2 *Preparing the Field and the Equations for the Program*

Examples have been given of mesh point temperature relationships. The reader should now be in a position to write further relationships for other situations which may arise.

In preparing the program, the first step is to define the field within an array of I rows and J columns, and to write the temperature equations in terms of (I, J) subscripting. It should be noted that because of differences in (I, J) subscripting, more than one equation may be required for a given physical situation, as for example, in four separate boundary corners. Figure 4.6 shows the field for a sample program, for part of a square duct, having 6 rows and 10 columns. Nine separate equations exist in the field, and of these equations 3, 4, 7 and 8 are used only once at the points shown, equation 2 is used for 8 points, equation 6 for 3 points, equations 5 and 9 for 4 points, and equation 1 for 22 points.

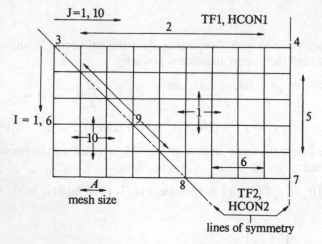

Fig. 4.6. *Field for an elementary BASIC program—part of a hollow duct. TF1, TF2, fluid temperatures; HCON 1, HCON 2, convection coefficients; TK, thermal conductivity.*

Equation (4.6) for the field is written in BASIC as

$$X = (T(I+1, J) + T(I-1, J) + T(I, J+1) + T(I, J-1))/4{\cdot}0 \quad (4.14)$$

where X is the temperature calculated in an iteration from previously calculated values. This is line 210 in the program, and together with the other nine equations, appears in the listing below.

BASIC Symbols and Iterative Sequence

```
50        CT1=HCON1*A*TF1/TK
51        CT2=HCON2*A*TF2/TK
52        D1=1.0+HCON1*A/TK
53        D2=2.0+HCON1*A/TK
54        D3=3.0+HCON2*A/TK
55        D4=2.0+HCON2*A/TK
60        ITER%=0
70        L%=0
80        FOR J%=1 TO 10
90        FOR I%=1 TO 6
100       K%=M%(I%,J%)
110       IF K%=1, GOTO 210
120       IF K%=2, GOTO 220
130       IF K%=3, GOTO 230
140       IF K%=4, GOTO 240
150       IF K%=5, GOTO 250
160       IF K%=6, GOTO 260
170       IF K%=7, GOTO 270
180       IF K%=8, GOTO 280
190       IF K%=9, GOTO 290
200       IF K%=10, GOTO 300
210       X=(T(I%+1,J%)+T(I%-1,J%)+T(I%,J%+1)+T(I%,J%-1))/4.0
215       GOTO 310
220       X=(0.5*(T(I%,J%+1)+T(I%,J%-1))+T(I%+1,J%)+CT1)/D2
225       GOTO 310
230       X=(T(I%,J%+1)+CT1)/D1
235       GOTO 310
240       X=(T(I%,J%-1)+T(I%+1,J%)+CT1)/D2
245       GOTO 310
250       X=(T(I%+1,J%)+T(I%-1,J%)+2.0*T(I%,J%-1))/4.0
255       GOTO 310
260       X=(0.5*(T(I%,J%+1)+T(I%,J%-1))+T(I%-1,J%)+CT2)/D4
265       GOTO 310
270       X=(T(I%,J%-1)+T(I%-1,J%)+CT2)/D4
275       GOTO 310
280       X=(T(I%,J%+1)+2.0*T(I%-1,J%)+CT2)/D3
285       GOTO 310
290       X=(T(I%-1,J%)+T(I%,J%+1))/2.0
295       GOTO 310
300       X=T(J%,I%)
310       DT=ABS(T(I%,J%)-X)
320       IF(DT>0.005)GOTO 340
330       L%=L%+1
340       T(I%,J%)=T(I%,J%)+1.9*(X-T(I%,J%))
350       NEXT I%
360       NEXT J%
370       ITER%=ITER%+1
380       IF(ITER%>200)GOTO 500
390       IF(L%<60)GOTO 70
400       CONTINUE
...
500       PRINT, "200 ITERATIONS REACHED"
```

4.2.3 The Iterative Technique

Starting with a set of given temperatures in the field, the iterative method consists of solving every equation in the field to obtain a new set of temperatures, each temperature being compared with the corresponding value in the previous iteration. If the difference in temperature is less than $0.005°$, then convergence is assumed at that point and iterations are continued until convergence has been obtained at all points. This method is known as the Gauss–Siedel iterative technique. The part of the program that carries out the selection of the correct equation at each point, and tests for convergence in the iteration, is reproduced on page 51. Initially, equation numbers are read into a storage array as integer values, $M(I, J)$. In lines 340–380 an accelerated convergence technique[3] is used. This is known as the extrapolated Liebmann method. The test for convergence at all points takes place in line 390, lines 400 and 500 being the continuation and end of the program, respectively.

Preparation of the complete program should now present no undue difficulties to readers conversant with BASIC. For further background material the reader is referred to Fenner[4].

The advantage of the method is that provided the field is correctly specified, and an appropriate set of equations is written, then any two-dimensional problem can be solved.

4.3 The Electrical Analogy of Conduction

The mathematical similarity between Fourier's law and Ohm's law has already been referred to. Thus, one-dimensional composite systems may be represented by a number of resistances in series, and it has been seen in Section 4.2 that two-dimensional fields may be represented by grids of resistances, as in Fig. 4.2 to Fig. 4.5. The technique may be extended to transient work by adding a capacitance at each node, the value being proportional to the thermal capacity of the node. Comparison of the two fundamental laws results in scaling factors being defined. Thus, let $Q = \theta/R_t$, and $I = V/R$, so

$$S_1 = \frac{I}{Q}; \quad S_2 = \frac{V}{\theta}; \quad S_3 = \frac{R}{R_t}$$

It is seen that values for only two of these can be chosen independently since $S_1 = S_2/S_3$. For transient work, two further scaling factors must be introduced. Thus $S_4 = T_e/T$, the ratio of electrical to thermal time constant, and $S_5 = C/C_t$, the ratio of electrical to thermal capacitance. Since the time constant is the product of resistance and capacitance it also follows that $S_4 = S_3 \times S_5$.

Complex two-dimensional shapes in steady state may be simulated by a continuous analogue using electrically conducting paper. Thus Fig. 4.7a shows a simple example with isothermal (constant voltage) boundary conditions. These are made using high-conductivity silver paint with an additional copper wire buried in the paint. Using the probe shown it is possible to plot the constant voltage lines between the boundaries.

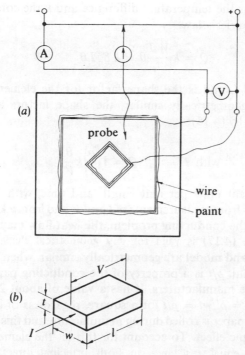

Fig. 4.7. Electrical analogy of two-dimensional conduction.

An element of conducting paper, length l, width w, and thickness t, is shown in Fig. 4.7b. There is a potential difference of V between the ends of the length l, and a current I is flowing.

If R is the resistance of the element, then $I = V/R$, and $R = \rho l/wt$, where ρ is the resistivity of the material; (units of ohms × length). Then,

$$I = \frac{wt}{\rho l} V = S_i \frac{t}{\rho} V \tag{4.15}$$

where S_i is the 'shape factor'. The shape factor for this rectangular element is w/l. Any other geometrical shape which passed the same current for the same voltage drop would have the same shape factor.

A similar equation may now be written for a geometrically similar element, length L, width W and thickness T, along which heat is conducting. θ is the temperature difference and k the conductivity, hence from Fourier's law,

$$Q = k \frac{WT}{L} \theta = k S_q T \theta \tag{4.16}$$

where $S_q = W/L$ and is the shape factor for the element. As the elements are geometrically similar, the shape factors are equal. Dividing (4.16) by (4.15):

$$\frac{Q}{I} = \frac{kT}{t/\rho} \frac{\theta}{V} \quad \text{with } T = 1 \quad Q = I \left(\frac{\rho}{t} \right) k \frac{\theta}{V} \tag{4.17}$$

Q will be in heat units per unit length and time, with T as unit thickness. Thus from measurements of I and V and from a knowledge of θ and k for the conducting problem, the heat flow may be calculated. Equation (4.17) is valid for any geometrical shape, provided the prototype and model are geometrically similar, when the shape factors are equal. ρ/t is a property of the conducting paper and is supplied by the manufacturers. It has a value of about 2000 ohms per square. ($R = \rho l/wt = \rho/t$ for a square, regardless of its size.)

Conducting paper is rolled during manufacture, and this can result in an anisotropic effect. To account for this let the element in Fig. 4.7b have an equal resistance in both principal directions. Then $\rho_1 l/wt = \rho_2 w/lt$, where ρ_1 is now the resistivity in the direction of current flow shown. It follows that $l/w = \sqrt{(\rho_2/\rho_1)}$. The model in Fig. 4.7$a$ would then be made rectangular to the extent indicated by measurements of ρ_1 and ρ_2.

Structures of composite material having differing thermal conductivities may be simulated by punching regular holes in the paper, to produce an area having an increased resistance. Convecting boundaries may be included either by cutting a band of paper outside the boundary into strips normal to the boundary, or by adding carbon resistors R_c at the edge[4], so that $R_c/R = (1/h)/(L/k\ WT)$ where h is the convection coefficient.

Three-dimensional shapes of rectangular form may be modelled using a number of sheets of paper to represent layers in the third coordinate direction, with additional resistances joining the centres of corresponding elements. Complicated three-dimensional shapes may be simulated in electrolytic tanks[5].

PROBLEMS

1. The diagram shows a plan view of the vertical insulation round the walls of a liquefied natural-gas storage tank. The inside and outside surface temperatures of the insulation are $-161°$ and $+1°C$. Calculate the heat transfer rate into the tank per metre height. Treat points a and b as fixed boundary temperatures of $-53°$ and $-107°C$ and assume one-dimensional conduction through the tank sides. Take k for the insulation as $50 \times 10^{-6}\ kW/(m\ K)$. The mesh size is 0·2 m. (Ans. 0·1764 kW/m.) (*The City University*).

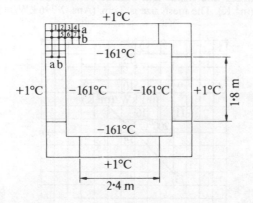

2. The diagram represents a square section with 81 mesh points. Boundary temperatures and convection coefficients are subscripted to allow for varying input information. Write a BASIC program to calculate temperatures at the 81 points. Note that there are only three separate physical situations: points 2, 3, 4 and 5 are similar, also points 6, 7, 8 and 9 are similar. Point 1 applies throughout the interior field. To check the program, the same convection coefficient on all faces will produce a symmetrical temperature field. Subsequent runs may be made with alternative boundary conditions.

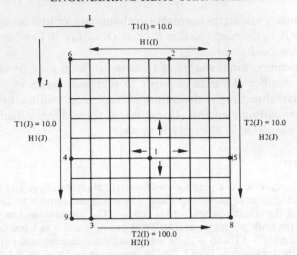

3. Write a BASIC program (deriving all the necessary equations) to calculate the heat transfer through the duct shown in the figure. The inside and outside fluid temperatures are 200° and 30°C, the thermal conductivity is 0·005 kW/(m K), and inside and outside convection coefficients are 0·1 and 0·05 kW/(m² K). The mesh size is 3 cm. (Ans. 2·796 kW/m.)

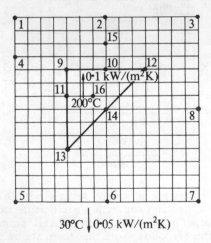

30°C $\downarrow$ 0·05 kW/(m²K)

4. An H-section copper conductor (see diagram) carries an overload current of 54,000 amps. In steady state conditions, the surface temperature is 60°C. Using a 0·5 cm grid, determine the temperatures within the copper. Calculate the total heat transfer at the surface, kW/cm length. The electrical resistivity

of copper is 2×10^{-8} ohm m, and the thermal conductivity is 0·381 kW/ (m K). (Ans. 0·73 kW/cm.)

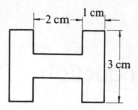

5. The figure shows a unique 1/8th part of a hollow square duct having inside and outside surface temperatures of 150°C and 15°C. Using the initial values of temperature given, relax until residuals have values not greater than ± 3. Calculate the heat transfer rate through the duct per m length by taking the average of the conduction rates at the 15°C, 70°C and 150°C isothermals. The thermal conductivity of the material of the duct is 4 W/m K. (Ans. 4240 W/m.)

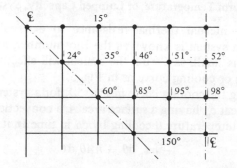

REFERENCES

1. Southwell, R. V. *Relaxation Methods in Theoretical Physics*, Oxford University Press (1946).
2. Bayley, F. J., Owen, J. M. and Turner, A. B. *Heat Transfer*, Nelson (1972).
3. McCracken, D. D. and Dorn, W. S. *Numerical Methods and Fortran Programming*, Wiley (1966).
4. Fenner, R. T. *Computing for Engineers.* Macmillan (1974).
4. Simonson, J. R. An Electrical Analogy of Extended Surfaces, *Bull. Mech. Engng. Educ.*, vol. 8, 215–25 (1969).
5. Karplus, W. J., and Soroka, W. W. *Analogue Methods*, McGraw-Hill Book Company, New York (1959).

5

Transient conduction

In any thermal system, transient heat transfer generally occurs before and after steady state operating conditions. The time duration of the transient condition can be of importance in design, and further, excessive thermal stress may arise. A simple approach is to assume the system is lumped, i.e. the temperature is uniform in space and is a function of time only. In a more detailed analysis, temperature will also be a function of position.

5.1 The Uniform Temperature, or Lumped Capacity, System

The ratio of internal thermal resistance to external convection resistance of a system is known as the Biot number, and when the Biot number is small, say <0.1, the system will effectively follow a single heating or cooling curve, as in Fig. 5.1.

Considering the cooling curve in Fig. 5.1, for a system of mass m, and specific heat c_p, having a surface area A, a convection coefficient h, and excess temperature θ cooling by $d\theta$ in time dt, it follows that

$$- mc_p \, d\theta = hA\theta \, dt$$

or

$$\frac{d\theta}{\theta} = - \frac{hA}{mc_p} dt = - \frac{dt}{T}$$

where $T = mc_p/hA$ = time constant = product of thermal resistance and capacitance.

Cooling from θ_1 to θ_2 will take time t given by

$$\ln \frac{\theta_2}{\theta_1} = - \frac{t}{T}$$

or

$$\theta_2/\theta_1 = e^{-t/T} \tag{5.1}$$

58

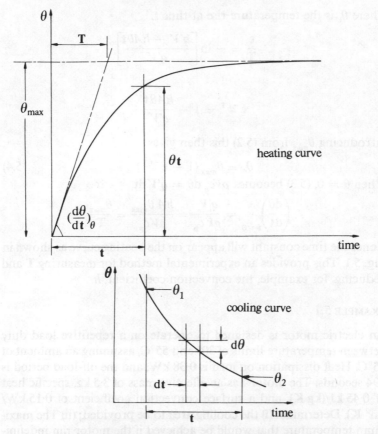

Fig. 5.1. Heating and cooling curves for lumped systems.

The heating curve in Fig. 5.1 is the result of an internal heat source, and the maximum temperature rise θ_{max} is achieved when the rate of cooling is equal to the rate of heat release. For a volume V and heat release rate q'

$$q'V = hA\theta_{max} \tag{5.2}$$

At some temperature θ the energy balance with a density ρ is given by

$$\rho V c_p \, d\theta = (q'V - hA\theta) \, dt$$

$$\therefore \frac{t}{\rho V c_p} = \left[-\frac{1}{hA} \ln (q'V - hA\theta) \right]_0^{\theta_t} \tag{5.3}$$

where θ_t is the temperature rise at time t.

$$\therefore \frac{t}{T} = - \ln\left[\frac{q'V - hA\theta t}{q'V}\right]$$

$$e^{-t/T} = 1 - \frac{hA\theta t}{q'V}$$

Introducing θ_{max} from (5.2) this then gives

$$\theta_t = \theta_{max}(1 - e^{-t/T}) \qquad (5.4)$$

When $\theta = 0$, (5.3) becomes $\rho V c_p \, d\theta = q'V \, dt$

$$\therefore \left(\frac{d\theta}{dt}\right)_{\theta=0} = \frac{q'V}{\rho V c_p} = \frac{hA \, \theta_{max}}{\rho V c_p} = \frac{\theta_{max}}{T}$$

Hence the time constant will appear on the heating curve as shown in Fig. 5.1. This provides an experimental method for measuring T and deducing, for example, the convection coefficient, h.

EXAMPLE 5.1

An electric motor is designed to operate on a repetitive load duty between temperature limits of 30° and 55°C, assuming an ambient of 15°C. Heat dissipation on load is 0·38 kW, and the off-load period is 294 seconds. The motor has an effective mass of 3·5 kg, specific heat of 0·45 kJ/(kg K), and a surface convection coefficient of 0·15 kW/ $(m^2 \text{ K})$. Determine: (i) the cooling area to be provided; (ii) The maximum temperature that would be achieved if the motor ran indefinitely; (iii) The duration of the allowable load period. (*The City University*)

Solution. The motor cools from 55° to 30°C in 294 seconds, hence the time constant may be found: $(30-15)/(55-15) = e^{-t/T}$ where $t = 294$ therefore $t/T = 0·982$ and $T = 300$ secs. But

$$T = mc_p/hA = (3·5 \times 0·45)/(0·15 \times A)$$

$$\therefore A = (3·5 \times 0·45)/(0·15 \times 300) = 0·035 \text{ m}^2$$

At the maximum temperature rise, all energy dissipation is convected away, hence $\theta_{max} hA = 0·38$

$$\therefore \theta_{max} = 0·38/(0·15 \times 0·035) = 72·5$$

The maximum temperature is therefore $72.5 + 15.0 = 87.5°C$.

To establish the load period, use equation (5.4) to determine the times to reach temperature rises of $40°$ and $15°$ from ambient. Thus:

$$40 = 72.5 (1 - e^{-t_1/300}) \therefore \mathbf{t_1} = 241 \text{ secs}$$

$$15 = 72.5 (1 - e^{-t_2/300}) \therefore \mathbf{t_2} = 69.6 \text{ secs}$$

The load period is $(\mathbf{t_1} - \mathbf{t_2}) = 171.4$ secs.

5.2 The Solution of Transient Conduction Problems in One Dimension

The discussion is limited to one-dimensional transient conduction in rectangular coordinates for which the equation is

$$\frac{\partial t}{\partial t} = \alpha \left(\frac{\partial^2 t}{\partial x^2} \right) \quad ((2.8))$$

For the general problem, numerical procedures will be described.

It is necessary to replace equation (2.8) by a finite difference relationship. Figure 5.2 shows a plane slab uniformly divided into sub-slabs of thickness a, with a temperature contour at some time t_0. Recalling the argument of section 4.1 it will be seen that the temperatures $t_{3,0}$, $t_{4,0}$, and $t_{5,0}$ are related

$$t_{3,0} + t_{5,0} = 2t_{4,0} + \left(\frac{\partial^2 t}{\partial x^2} \right)_{4,0} a^2$$

and hence

$$\left(\frac{\partial^2 t}{\partial x^2} \right)_{4,0} = \frac{t_{3,0} + t_{5,0} - 2t_{4,0}}{a^2} \quad (5.5)$$

With a *forward* time step, the finite difference relationship for $(\partial t / \partial t)$ is

$$\left(\frac{\partial t}{\partial t} \right)_{4,0} = \frac{t_{4,1} - t_{4,0}}{\Delta t}$$

where $t_{4,1}$ is the temperature at point 4 at time $\mathbf{t_1}$, which is Δt after $\mathbf{t_0}$. Equation (2.8) can now be replaced by

$$\frac{t_{4,1} - t_{4,0}}{\Delta t} = \alpha \left(\frac{t_{3,0} + t_{5,0} - 2t_{4,0}}{a^2} \right) \quad (5.6)$$

This may be re-arranged as

$$t_{4,1} = F(t_{3,0} + t_{5,0}) + t_{4,0}(1 - 2F) \quad (5.7)$$

where F is the Fourier number, $\Delta t \alpha/a^2 = \Delta t k/\rho c_p a^2$. This compares energy conducted in time Δt, proportional to $\Delta t\, k/a$, to energy stored, proportional to $\rho c_p a$, and hence gives a measure of temperature response.

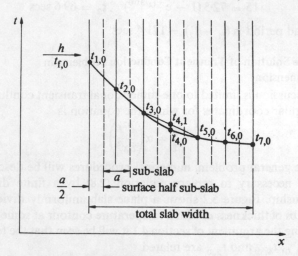

Fig. 5.2. *Treatment of a plane slab for transient conduction in one dimension. (First subscript denotes position, second subscript denotes time.)*

5.2.1 A Numerical Method

Equation 5.7 is suitable for performing a numerical solution. Values of Δt and a are chosen to give a suitable value of F. The coefficient of $t_{4,0}$ must remain positive for the solution to be stable, hence $F \leqslant \frac{1}{2}$. If $F = \frac{1}{2}$ then $t_{4,1}$ is simply the average of $t_{3,0}$ and $t_{5,0}$. For smaller values of F increased accuracy will be obtained. In order to solve a problem it is, of course, necessary to know the boundary temperatures after each time interval Δt.

EXAMPLE 5.2

The surfaces of a brick wall, 300 mm thick initially 20°C throughout, rise in temperature at a constant rate of 10°C every 2500 seconds. Dividing the wall into six equal slabs find the temperature distribution in the wall after 10^4 seconds. Use $F = \frac{1}{2}$. $\alpha = 0.05 \times 10^{-5}$ m²/sec.

Solution. It is seen that $F = \Delta t\, \alpha/a^2 = 2500 \times 0{\cdot}05/(10^5 \times 0{\cdot}05^2) = 0{\cdot}5$. From equation (5.7), with $F = \frac{1}{2}$, the following table may be written:

Node	Surface	1	2	3	4	5	Surface
Initial temperatures	20	20	20	20	20	20	20
At: 2500 secs	30	20	20	20	20	20	30
5000 secs	40	25	20	20	20	25	40
7500 secs	50	30	22·5	20	22·5	30	50
10000 secs	60	36·25	25	22·5	25	36·25	60

Convection at a solid boundary can be allowed for in a numerical solution by setting up an energy equation for the boundary slab. The change in stored energy over the chosen time interval is equal to the conduction from the adjoining slab plus the convection from the boundary fluid. A relation between the fluid temperature and temperatures in the solid is thus obtained.

This procedure is based on the electrical analogy of transient conduction in which a capacitance is added at each node. Referring to the surface half-sub-slab in Fig. 5.2

$$\frac{(a \times 1)}{2} \times \rho \times c_p\,(t_{1,1} - t_{1,0}) = \frac{k \times 1(t_{2,0} - t_{1,0})\,\Delta t}{a}$$
$$+\; h \times 1(t_{f,0} - t_{1,0})\,\Delta t$$

$$\therefore t_{1,1} = \left(\frac{2k\,\Delta t}{\rho c_p a^2}\right) t_{2,0} + \left(\frac{2h\,\Delta t}{\rho c_p a}\right) t_{f,0} + t_{1,0}\left(1 - \frac{2k\,\Delta t}{\rho c_p a^2} - \frac{2h\,\Delta t}{\rho c_p a}\right)$$

$$\therefore t_{1,1} = 2F\, t_{2,0} + 2FB\, t_{f,0} + t_{1,0}\,(1 - 2F(1 + B)) \qquad (5.8)$$

where $B = ha/k$, which is the Biot number of a sub-slab, and F is the Fourier number as before. A new stability criterion applies to the boundary equation, i.e., $F(1 + B) \leqslant \frac{1}{2}$, for the coefficient of $t_{1,0}$ to remain positive. This means that $F < \frac{1}{2}$ and the numerical procedure using equations (5.7) and (5.8) is more complicated throughout the whole field.

5.2.2 BASIC Program for Transients in One Dimension with Convection at the Boundary

A one-dimensional problem with convecting boundary conditions can readily be solved for as many time steps as are required by a BASIC program. Equation (5.8) is used at the two boundaries and equation

(5.7) is used at all points in between. Temperatures are not normally stored beyond one time step, hence they may be printed at chosen intervals throughout the transient. A program is given, by way of example, to solve the following problem.

EXAMPLE 5.3

It is required to investigate temperature rise and heat penetration in a fire door. The door is 50 mm thick, and is considered as ten slabs 5 mm thick (hence there are two boundary temperatures from equation (5.8), and nine interior temperatures from equation (5.7).) Initially the temperature of the door and environment on both sides may be taken as 25°C. On one side of the door a fire causes the effective environment temperature to rise linearly to 500°C in 200 seconds, and then to remain constant at that temperature, with a combined convection and radiation coefficient to the door of 120 W/m² K. On the other side, the environment temperature remains constant at 25°C, and a convection coefficient of 15 W/m² K applies. The problem is shown in the figure.

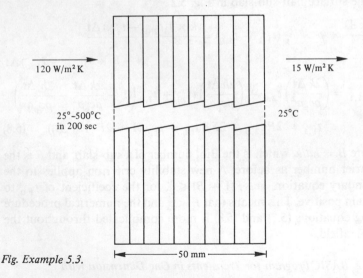

Fig. Example 5.3.

Determine the temperature distribution throughout the door every 200 sec for a total time of 1800 sec, the heat flow rates to and from the door and the energy stored in the door at these times.

For the door, $k = 0.19$ W/m K, $\rho = 577$ kg/m³, and $C_p = 816$ J/kg K.

Solution. The program listing and results are given below. Heat flow and heat stored results are all per m² of door, and they are calculated in lines 500, 510 and 540. It is seen that initially heat transfer rates to the door are high, and the rate falls as the outer surface of the door rises in temperature. The temperature of the inner face of the door rises slowly, however by 1800 sec the temperature is significant and the heat flow rate is approaching 1 kW/m²; but the initial slow response and corresponding escape time for occupants are clearly apparent.

BASIC Program Listing

```
10          DIMENSION T(11),TNEW(11)
20          A=0.005
30          TK=0.19
40          DEN=577.0
50          CP=816.0
60          H1=120.0
70          H2=15.0
80          T1=25.0
90          T2=25.0
100         TTOP=500.0
110         TRT=200.0
120         TLIM=1800.0
130         TPRINT=200.0
140         DTPRINT=200.0
150         B1=H1*A/TK
160         B2=H2*A/TK
170         F=TK/(DEN*CP*A*A)
180         TIME1=1.0/(2.0*F*(1.0+B1))
190         TIME2=1.0/(2.0*F*(1.0+B2))
200         IF(TIME1>TIME2)GO TO 230
210         TIME=TIME1
220         GO TO 240
230         TIME=TIME2
240         PRINT,"TIME STEP IS:",TIME
250         PRINT,"DO YOU WISH TO ROUND OFF AT A SMALLER VALUE OF TIME STEP?"
260         PRINT,"1 - YES; 2 - NO"
270         INPUT IX%
280         IF(IX%>1)GO TO 300
290         INPUT TIME
300         F=F*TIME
310         F2=2.0*F
320         FB12=2.0*F*B1
330         FB22=2.0*F*B2
340         FT1=1.0-2.0*F*(1.0+B1)
350         FT2=1.0-2.0*F*(1.0+B2)
360         FT=1.0-2.0*F
370         DTR=(500.0-25.0)*TIME/TRT
380         TRUN=0.0
390         T1=T1+DTR
395         IF(T1>TTOP)THEN T1=TTOP
400         TNEW(1)=F2*T(2)+FB12*T1+T(1)*FT1
410         FOR I=2,10
420         TNEW(I%)=F*(T(I%-1)+T(I%+1))+T(I%)*FT
430         NEXT I%
440         TNEW(11)=F2*T(10)+FB22*T2+T(11)*FT2
450         TRUN=TRUN+TIME
460         FOR I%=1,11
470         T(I%)=TNEW(I%)
480         NEXT I%
490         IF(TRUN<TPRINT)GO TO 390
500         HEATIN=(T1-T(1))*H1
510         HEATOUT=(T(11)-T2)*H2
```

```
520        HSTORE=0.0
530        FOR I%=1,11
540        HSTORE=HSTORE+(T(I%)-25.0)*A*DEN*CP
550        NEXT I%
560        PRINT,"HEAT TO DOOR =",HEATIN," WATTS"
561        PRINT,"HEAT FROM DOOR =",HEATOUT," WATTS"
562        PRINT,"HEAT STORED =",HSTORE," JOULES"
565        PRINT,"TEMPERARURES -"
570        PRINT T(I%) FOR I%=1,11
580        TPRINT=TPRINT+DTPRINT
590        IF(TPRINT>TLIM)GO TO 620
600        GO TO 370
620        STOP
```

Results from Listing

```
TIME STEP IS:  0.74498734E 01
DO YOU WISH TO ROUND OFF AT A SMALLER VALUE OF TIME STEP?
1 - YES; 2 - NO
=1
=5.0
TIME =   200.0 SECONDS
HEAT TO DOOR =      8694.08 WATTS
HEAT FROM DOOR =        0.16 WATTS
HEAT STORED = 1791252.77 JOULES
TEMPERATURES -
427.5 227.9 118.6  64.4  40.0  30.2  26.6  25.5  25.1  25.0  25.0
TIME =   400.0 SECONDS
HEAT TO DOOR =      4539.30 WATTS
HEAT FROM DOOR =       17.69 WATTS
HEAT STORED = 3005670.75 JOULES
TEMPERATURES -
462.2 344.6 242.3 162.2 105.4  68.9  47.3  35.6  29.8  27.1  26.2
TIME =   600.0 SECONDS
HEAT TO DOOR =      3531.62 WATTS
HEAT FROM DOOR =      110.10 WATTS
HEAT STORED = 3808477.53 JOULES
TEMPERATURES -
470.6 378.5 293.7 220.2 160.4 114.6  81.6  59.2  45.0  36.6  32.3
TIME =   800.0 SECONDS
HEAT TO DOOR =      2994.03 WATTS
HEAT FROM DOOR =      266.80 WATTS
HEAT STORED = 4438335.19 JOULES
TEMPERATURES -
475.0 396.8 323.0 256.3 198.8 151.3 113.9  85.6  65.4  51.6  42.8
TIME = 1000.0 SECONDS
HEAT TO DOOR =      2645.02 WATTS
HEAT FROM DOOR =      439.52 WATTS
HEAT STORED = 4948213.50 JOULES
TEMPERATURES -
478.0 408.7 342.6 281.5 227.1 180.2 141.2 110.0  85.8  67.6  54.3
TIME = 1200.0 SECONDS
HEAT TO DOOR =      2394.35 WATTS
HEAT FROM DOOR =      600.89 WATTS
HEAT STORED = 5362725.00 JOULES
TEMPERATURES -
480.0 417.3 356.9 300.3 248.8 203.2 163.9 131.0 104.1  82.4  65.1
TIME = 1400.0 SECONDS
HEAT TO DOOR =      2203.54 WATTS
HEAT FROM DOOR =      741.44 WATTS
HEAT STORED = 5700336.75 JOULES
TEMPERATURES -
481.6 423.9 367.9 314.9 265.9 221.7 182.6 148.7 119.8  95.3  74.4
TIME = 1600.0 SECONDS
HEAT TO DOOR =      2053.42 WATTS
HEAT FROM DOOR =      859.95 WATTS
```

```
HEAT STORED = 5976912.13 JOULES
TEMPERATURES -
482.9 429.0 376.6 326.5 279.7 236.7 197.9 163.3 132.9 106.1  82.3
TIME = 1800.0 SECONDS
HEAT TO DOOR =     1933.17 WATTS
HEAT FROM DOOR =    958.35 WATTS
HEAT STORED = 6202445.25 JOULES
TEMPERATURES -
483.9 433.2 383.6 335.9 290.9 248.9 210.4 175.4 143.7 115.0  88.9
```

It may be seen from Fig. 5.2 that the averaging of temperatures can be carried out by drawing. This is the basis of the Binder–Schmidt method[1,2]. Graphical constructions are also possible for convecting boundaries. These rely on the fact that the conduction rate at the boundary must equate to the convection rate. Various cases are considered by Hsu[3].

5.3. Two-dimensional Transient Conduction

In two-dimensional transient conduction in rectangular coordinates the differential equation is

$$\frac{\partial t}{\partial t} = \alpha\left(\frac{\partial^2 t}{\partial x^2} + \frac{\partial^2 t}{\partial y^2}\right) \tag{5.9}$$

and referring back to the nomenclature of Fig. 4.1, the finite difference relationship for a field point with a forward time step can be seen to be

$$\frac{t_{0,1} - t_{0,0}}{\Delta t} = \alpha\left(\frac{t_{1,0} + t_{2,0} + t_{3,0} + t_{4,0} - 4t_{0,0}}{a^2}\right) \tag{5.10}$$

This is re-arranged to give

$$t_{0,1} = F(t_{1,0} + t_{2,0} + t_{3,0} + t_{4,0}) + t_{0,0}(1 - 4F) \tag{5.11}$$

with the stability requirement that $F \leqslant \frac{1}{4}$.

In transient work, alternate finite difference relationships having backward time steps may be used. The equivalent of equation (5.10) would be:

$$\frac{t_{0,1} - t_{0,0}}{\Delta t} = \alpha\left(\frac{t_{1,1} + t_{2,1} + t_{3,1} + t_{4,1} - 4t_{0,1}}{a^2}\right) \tag{5.12}$$

Then $t_{0,0}$ is the only known temperature, and equations for all points must be solved simultaneously to obtain the temperatures after the next time step. However, there is no stability restriction in this case.

The reader is referred to Bayley[4] for a full discussion of these methods.

The two-dimensional steady state computing methods discussed in Section 4.2.2 may be modified to deal with transient problems using equations with forward time steps, by substituting transient equations for the steady state ones and by replacing the iterative technique by a scheme for solving the equations throughout the field for as many time steps as are required. Boundary equations may be derived following a similar procedure to that in Section 5.2.1. To illustrate, it may be verified that the transient equation for point 2 in Fig. 4.6 is, in BASIC

$$T(I, J) = F*(TP(I, J + 1) + TP(I, J - I) + 2·0*TP(I + 1, J))$$

$$+ 2·0*F*B1*TF1 + TP(I, J)*S(2)$$

where $T(I, J)$ denotes the new temperature, $TP(I, J)$ the existing temperature, and

  TK = thermal conductivity of material
  DT = time interval
  D   = density of material
  C   = specific heat of material
  A   = mesh size
  $F = TK*DT/(D*C*A**2)$
  $S(2) = (1·0-4·0*F-2·0*F*B1)$
  $B1 = HCON1*A/TK$

Since different stability criteria exist the coefficient $S(I)$ is subscripted to enable a DT value to be determined to satisfy all equations.

5.4 Periodic Temperature Changes at a Surface

A periodically changing surface temperature can also be dealt with by numerical or graphical methods, but the work involved is probably not justified in view of the fact that an analytical solution is not too lengthy for this particular boundary condition.

The problem to be considered is one in which a plane slab of material, referred to as a 'semi-infinite solid', is regarded as being infinitely thick, the periodic surface temperature existing at the face of the slab where $x = 0$. The surface temperature varies in a sinusoidal manner and, because of the assumption of infinite thickness, the temperature history within the material is controlled only by the surface variation. Further, conduction takes place in only one dimension, so that edge effects are neglected or the specimen is regarded as being sufficiently large in the y-direction for conduction to be one-dimensional over the area of material of interest. An additional assumption is that the cyclic variation of temperature at the surface has been going on for a time sufficiently long for temperatures elsewhere in the slab to be repeated identically in each cycle. The general result obtained, as will be seen, is that the interior temperature cycle lags behind the surface variation, depending on the depth and, in addition, has a diminished amplitude compared with the maximum surface values. This type of analysis finds application wherever a cyclic variation of temperature occurs, as in annual or daily temperature variation of buildings or the ground exposed to solar radiation, and, in the other extreme, in the cylinders of reciprocating engines. The chief restriction on the validity of the analysis is whether the object in question may be regarded as infinitely thick. The depth in the material at which the temperature amplitude has become, say, only 1 per cent of the surface value is the criterion by which this is judged.

The surface of the slab has a mean temperature t. It varies in a sinusoidal manner between an upper temperature limit of $t + (\theta_m)_0$ and a lower limit of $t - (\theta_m)_0$. Thus, if θ is the temperature difference between the actual temperature at any instant and the mean then, at the surface where $x = 0$, θ varies between $\pm (\theta_m)_0$ where θ_m denotes the maximum difference. Further, at some depth x in the slab, θ varies between $\pm (\theta_m)_x$. The frequency of the temperature variation is n cycles per unit time, so $1/n$ is the period of the variation. The boundary conditions of the problem are set by the sinusoidal temperature variation at the surface, given by

$$\theta = (\theta_m)_0 \sin(2\pi n t) \tag{5.13}$$

which is the value of θ at $x = 0$, and $t = t$. At $x = 0$ and $t = 0$, $\theta = 0$. In equation (5.13) $2\pi n$ is the angular velocity of the sine wave in rads/unit time.

Since θ is the temperature variation about a mean value t, θ may be regarded as the temperature variable since t is constant. For this case the one-dimensional unsteady equation, (2.8), becomes

$$\frac{\partial \theta}{\partial t} = \alpha \left(\frac{\partial^2 \theta}{\partial x^2} \right) \tag{5.14}$$

Since θ varies sinusoidally at the surface, it can also be expected to do so within the solid, but between reducing limits and further, the phase shift will depend on the time to penetrate to depth x, hence the form of solution chosen is

$$\theta = Ce^{-px}\sin(2\pi nt - qx) \qquad (5.15)$$

where C, p, and q are constants to be determined. The constants p and q may be found by substituting equation (5.15) in equation (5.14). The partial differential coefficients found from (5.15) are

$$\frac{\partial\theta}{\partial t} = 2\pi nCe^{-px}\cos(2\pi nt - qx)$$

$$\frac{\partial\theta}{\partial x} = -pCe^{-px}\sin(2\pi nt - qx) - qCe^{-px}\cos(2\pi nt - qx)$$

$$\frac{\partial^2\theta}{\partial x^2} = p^2Ce^{-px}\sin(2\pi nt - qx) + pqCe^{-px}\cos(2\pi nt - qx)$$
$$+ pqCe^{-px}\cos(2\pi nt - qx) - q^2Ce^{-px}\sin(2\pi nt - qx)$$

Hence, equation (5.14) becomes, noting that Ce^{-px} may be cancelled from all terms:

$$2\pi n\cos(2\pi nt - qx) = \alpha[p^2\sin(2\pi nt - qx)$$
$$+ 2pq\cos(2\pi nt - qx) - q^2\sin(2\pi nt - qx)]$$

$$\therefore \quad (2\pi n - 2pq\alpha)\cos(2\pi nt - qx) = \alpha(p^2 - q^2)\sin(2\pi nt - qx)$$

Since there is no cosine term on the right, it follows that

$$(2\pi n - 2pq\alpha) = 0$$

$$\therefore \quad pq = \pi n/\alpha$$

Further, as there is no sine term on the left, it follows that

$$p^2 - q^2 = 0$$

$$\text{or} \quad p = q$$

Thus, from these results,

$$p = q = \pm(\pi n/\alpha)^{0.5}$$

The negative solution means an exponential increase of θ with x, hence taking the positive result only, equation (5.15) becomes:

$$\theta = C\exp[-x(\pi n/\alpha)^{0.5}]\sin[2\pi nt - x(\pi n/\alpha)^{0.5}] \qquad (5.16)$$

This result may now be compared with the boundary condition at $x = 0$ and $t = t$. Thus (5.16) gives

$$\theta = C \sin 2\pi n t$$

and the boundary condition gives

$$\theta = (\theta_m)_0 \sin 2\pi n t$$

Thus comparing these two equations shows that $C = (\theta_m)_0$. The final solution is therefore

$$\theta = (\theta_m)_0 \exp[-x(\pi n/\alpha)^{0.5}] \sin[2\pi n t - x(\pi n/\alpha)^{0.5}] \qquad (5.17)$$

This equation shows that the maximum variation of θ decreases exponentially with x, the distance into the solid, according to the equation

$$(\theta_m)_x = (\theta_m)_0 \exp[-x(\pi n/\alpha)^{0.5}] \qquad (5.18)$$

The general form of the result given by equation (5.17) is shown in Figs. 5.3 and 5.4. In Fig. 5.3 the temperature variation with distance

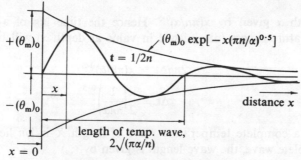

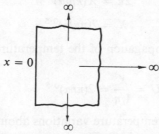

Fig. 5.3. Established temperature variation vs. distance into solid, at $t = 1/2n$.

at a chosen time is shown, and in Fig. 5.4 temperature variations with time at the surface and depth x are shown. It will be seen that a temperature wave propagates into the solid, and also that the cyclic variation of temperature at some depth x lags behind the surface variation. The phase difference in temperature variation

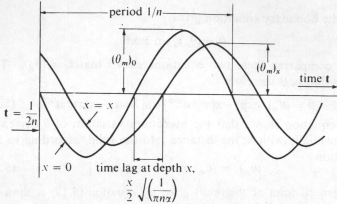

Fig. 5.4. *Temperature variation with time at $x = 0$, and at depth x.*

at depth x given by $x(\pi n/\alpha)^{0.5}$. Hence the time lag of a certain temperature excess (diminished in value at depth x) will be given by

$$2\pi n \Delta t = x(\pi n/\alpha)^{0.5}$$

$$\therefore \quad \Delta t = \frac{x}{2}\left(\frac{1}{\pi n\alpha}\right)^{0.5} \tag{5.19}$$

Δt for a complete temperature wave of length X is $1/n$ hence, for a complete wave, the wave-length is given by

$$2\pi = X(\pi n/\alpha)^{0.5}$$

$$\therefore \quad X = 2(\pi\alpha/n)^{0.5} \tag{5.20}$$

Also, the velocity of propagation of the temperature wave into the solid is

$$U = \frac{X}{1/n} = 2(\pi n\alpha)^{0.5} \tag{5.21}$$

The ratio of maximum temperature variations about the mean may be compared using equation (5.18). Thus:

$$\frac{(\theta_m)_x}{(\theta_m)_0} = \exp[-x(\pi n/\alpha)^{0.5}]$$

If it is required to determine the distance x at which $(\theta_m)_x$ has decreased to a certain percentage of $(\theta_m)_0$, this result may be re-arranged to give

$$x = \frac{\ln\left[(\theta_m)_0/(\theta_m)_x\right]}{(\pi n/\alpha)^{0.5}} \qquad (5.22)$$

Finally, the heat transfer rate at the wall surface, at $x = 0$, may be determined from

$$q = \frac{Q}{A} = -k\left(\frac{\partial\theta}{\partial x}\right)_{x=0}$$

and from equation (5.17),

$$\left(\frac{\partial\theta}{\partial x}\right)_{x=0} = -(\theta_m)_0(\pi n/\alpha)^{0.5}(\sin 2\pi nt + \cos 2\pi nt)$$

Using the identity,

$$\sin(2\pi nt + \pi/4) = \sin 2\pi nt \cos \pi/4 + \sin \pi/4 \cos 2\pi nt$$

$$= (1/\sqrt{2})(\sin 2\pi nt + \cos 2\pi nt)$$

it follows that

$$\left(\frac{\partial\theta}{\partial x}\right)_{x=0} = -(\theta_m)_0(2\pi n/\alpha)^{0.5} \sin(2\pi nt + \pi/4)$$

and hence,

$$q = k(\theta_m)_0(2\pi n/\alpha)^{0.5} \sin(2\pi nt + \pi/4) \qquad (5.23)$$

From equation (5.23) it will be seen that the surface heat transfer rate varies sinusoidally and with the same frequency as the surface temperature, but leading by a period of $1/8n$. The total heat transfer at the wall is given by

$$\int q \, dt = \int k(\theta_m)_0(2\pi n/\alpha)^{0.5} \sin(2\pi nt + \pi/4) \, dt$$

$$= -k(\theta_m)_0(1/2\pi n\alpha)^{0.5} \cos(2\pi nt + \pi/4) \text{ heat/unit area} \qquad (5.24)$$

Thus the energy stored, as represented by an integral of heat transfer rate at the surface, also varies with the same frequency, but it will be found that it lags behind the surface temperature variation by a period of $1/8n$. Further, it will be apparent that the surface heat transfer is both to and from the solid and that the energy stored is in sequence both positive and negative relative to the mean temperature.

EXAMPLE 5.4

In a cyclic heating process the inside of a furnace wall is subjected to a sinusoidal temperature variation. The temperature rises from a minimum of 100°C to a maximum of 750°C in 3 hours. With $k = 692 \times 10^{-6}$ kW/(m K) and $\alpha = 0.0516 \times 10^{-5}$ m²/sec, determine: (i) the velocity of the temperature wave penetrating the wall; (ii) the time lag of the wave function at a depth of 0.2 m compared with the surface; (iii) the maximum and minimum temperatures at a depth of 0.2 m; (iv) the temperature at a depth of 0.2 m when the surface temperature is a maximum; and the surface temperature when the temperature at a depth of 0.2 m is a maximum (*The City University*).

Solution. (i) The period $(1/n)$ is 6 hours. From (5.21)

$$U = 2(\pi \times 0.0516 \times 3600/6 \times 10^5)^{0.5}$$
$$= 0.0622 \text{ m/h}$$

(ii) The time lag is given by (5.19),

$$\Delta t = \frac{0.2}{2}\left(\frac{6 \times 10^5}{\pi \times 0.0516 \times 3600}\right)^{0.5}$$

$$= 3.22 \text{ hours}$$

(Check: 0.2 m at 0.0622 m/h takes 3.22 hours.)

(iii) The maximum and minimum temperatures at 0.2 m are obtained from (5.18). $(\theta_m)_0$ is 325

$$(\theta_m)_x = 325 \times \exp\left[-0.2(\pi \times 10^5/6 \times 0.0516 \times 3600)^{0.5}\right]$$

$$= 325 \times 0.03477 = 11.3$$

The mean temperature is 425°C, hence the maximum is 436.3°C, and the minimum is 413.7°C.

(iv) Using equation (5.17); $\theta = 325 \times 0.03477 \sin(2\pi nt - 1.07\pi)$. Surface temperature is a maximum at $t = 1\frac{1}{2}$ hours $= \frac{1}{4}n$ (mean to maximum), hence $\theta = 11.3 \sin(0.5\pi - 1.07\pi) = 11.3 \sin(-0.57\pi) = -11.3 \sin 0.43\pi = -11.02°$. Hence temperature at 0.2 m is $425 - 11.02 = 414°C$. The temperature wave at 0.2 m must advance in phase by 1.07π to reach its maximum value, when the surface wave will be at 1.57π. Hence $\theta = 325 \sin(1.57\pi) = -325 \sin(0.43\pi) = -317°$

∴ the surface temperature is $425 - 317 = 108°C$

PROBLEMS

1. Steel strip of thickness 1·27 cm emerges from a rolling mill at a temperature of 538°C and with a velocity of 2·44 m/sec. The strip is cooled in such a way that its surface temperature falls linearly with distance from the mill at a rate of 110°C/m.

Derive a finite difference method for dealing with this case of transient heat conduction assuming that heat flows only in the direction normal to the strip faces.

Subdividing the strip into six increments of thickness, determine the temperature distribution in the strip and the heat flux from the surface at a position 2·74 m from the mill. (For steel take thermal conductivity 43·3 × 10⁻³ kW/(m K), thermal diffusivity 0·98 × 10⁻⁵ m²/s.) (Ans. 3·41 × 10³ kW/(m²)(*University of Manchester*).

2. A steel pipe, 2·54 cm wall thickness, is initially at a uniform temperature of 16°C when a liquid metal at 572°C is pumped through it for a time of 10 sec and with a surface coefficient of 2·84 kW/(m² K). It may be assumed that the pipe diameter is large enough for the wall to be considered plane, that no heat loss occurs from the outside of the pipe and from the inside after the flow of liquid metal has ceased.

Derive a numerical method to deal with this case using finite increments of thickness and making the simplification that the heat capacity of the surface half-increment is negligible. Using four increments determine the wall temperature distribution after 18 sec. (For steel take thermal conductivity 0·041 kW/(m K), density 7530 kg/m³ and specific heat 0·536 kJ/(kg K.) (Ans. 231°C, 120°C.) (*University of Manchester*).

3. At a certain instant in transient one-dimensional conduction through a 3 cm thick slab of chrome steel, 40 cm square, the temperature distribution along the 3 cm thickness is given by $t = (60 + 1.2x^2 + 0.3x^3)°C$ where x is the distance from one 40 cm square face. Calculate the rate of energy storage in the slab, and the rate of change of temperature at each square face, at the particular instant. Take $\rho = 7833$ kg/m³, $k = 0.0398$ kW/(m K), $c_p = 0.46$ kJ/(kg K). (Ans. +9·74 kW, +0·266 K/s and +0·865 K/s.) (*The City University*).

4. Given the differential equation $(\partial t/\partial \tau) = \alpha(\partial^2 t/\partial x^2)$, for unsteady conduction in a 'one-dimensional' wall, show that the temperature $t_{n, p+1}$ at some section n and time instant $(p + 1)$ can be calculated approximately from

$$t_{n, p+1} = F\left[t_{n+1, p} + t_{n-1, p} + \left(\frac{1}{F} - 2\right)t_{n, p}\right]$$

The temperatures in the right-hand bracket are values at equidistant sections $(n - 1), n, (n + 1)$, preceding $(p + 1)$ by a finite time interval $\Delta\tau$; $F = \alpha\Delta\tau/\Delta x^2$ is the Fourier number.

Plane 1 is a distance $\Delta x/2$ to the right of a wall surface. Prove that, if the convection coefficient and temperature of the fluid to the left of the surface is h and t_s respectively,

$$t_{1, p+1} = F\left[t_{2, p} + \frac{2B}{2 + B}t_{s, p} + \left(\frac{1}{F} - \frac{2 + 3B}{2 + B}\right)t_{1, p}\right]$$

where $B = h\Delta x/k$. (*University of Bristol*).

5. An insulating screen is intended to withstand the penetration of high temperature for as long as possible. Select either material A, B, or C as being best for this purpose.

Material	k kW/(m K)	ρ kg/m^3	c_p kJ/(kg K)
A	600×10^{-6}	1500	0·84
B	600×10^{-6}	1200	1·60
C	280×10^{-6}	750	1·10

The screen is 5·3 cm thick and is divided into 5 increments. It is initially 15°C throughout; the temperature of one face rises linearly by 20°C per minute. Regard the other face as insulated. Determine, for the chosen material, and by a numerical technique, the time before the insulated face temperature starts to rise, and the temperature of this face after 18 minutes. (Ans. B, 15 mins, 24·4°C.) (*The City University*).

6. In order to carry out an approximate analysis of a butt-welding process, it is assumed that there is a uniform rate of heat generation at the contact face between the two bars, that heat conduction occurs only in a direction normal to the contact face and that the physical properties of the bars are constant. Derive a numerical method to deal with this case by sub-dividing the bars into finite increments of length.

Apply the method to obtain the approximate temperature distribution in two similar steel bars, initially at 16°C, after 10 sec. The heat generation rate is 1·005 kJ/(cm^{2}s) and this acts for a period of 5 sec. Use 0·635 cm increments of length and, for steel, take thermal conductivity $= 45 \times 10^{-3}$ kW/(m K), density $= 7690$ kg/m^3 and specific heat $= 0·545$ kJ/(kg K). (Ans. 1249°C maximum at joint, after 10 sec: 534°C.) (*University of Manchester*).

7. A slab of material is considered as six layers of equal thickness, and is initially at $20°C$ throughout. The two surfaces of the slab rise in temperature by $10°C$ per time interval when a transient analysis is undertaken with $F = \frac{1}{2}$. Calculate the centre-line temperature of the slab when the surface temperature is $60°C$. Repeat the calculation for $F = \frac{1}{4}$ to give a more accurate solution, and find the percentage error in the fresh answer. (Ans. $22·5°C$, $23·32°C$, $3·52$ per cent.)

8. A large steel plate $7·62$ cm thick initially uniformly at $816°C$ is quenched in oil at $38°C$. If the oil temperature remains constant and there is negligible surface resistance, estimate the time required to reduce to $427°C$:
  (a) the average temperature of the slab,
  (b) the centre-line temperature.
Thermal diffusivity of steel $= 1·032 \times 10^{-5}$ m^2/s. (Ans. $26·7$ sec, $52·0$ sec.) (*University of Leeds*).

9. A current of 3 amps is passed along a 1 mm diameter wire of resistance $3·5$ ohms/m. The wire reaches a steady temperature of $60°C$ in an atmosphere of $20°C$. Calculate the initial rate of temperature rise of the wire, and the temperature after a time lapse equal to the thermal time constant of the wire. The mass is 25 g/m and the specific heat $0·460$ kJ/(kg K). (Ans. $2·74$ K/s, $45·2°C$.) (*The City University*).

10. An internal combustion engine runs at 2500 r.p.m. The thermal diffusivity of the carbon steel of the cylinder walls is $1·16 \times 10^{-5}$ m^2/s. The temperature of the cylinder wall varies sinusoidally between 5000 and $100°C$. Assuming that the cylinder wall behaves like a semi-infinite solid, determine the depth into the wall in cm at which the temperature amplitude has decreased to 1 per cent of the surface value, and plot the heat transfer rate at the wall surface over a complete cycle. $k = 40 \times 10^{-3}$ kW/(mk). (Ans. $0·194$ cm, limits $\pm 2·68 \times 10^4$ kW/m^2.)

11. Write a transient program in BASIC for the section in Question 3, Chapter 4. Assume the duct is initially at $30°C$ throughout, and then gas at $200°C$ enters the duct. Find the temperature distribution and the total heat transfer into and out of the duct at 30 and at 360 seconds after the hot gas enters. (Ans. 30 secs: 279 kJ in, 78×10^{-6} kJ out; 360 secs: 2304 kJ in, $18·95$ kJ out.)

REFERENCES

1. Binder, L. *Dissertation*, München (1910).
2. Schmidt, E. *Festschrift zum siebzigsten Geburtstag August Föppl*, Springer, Berlin (1924).
3. Hsu, S. T. *Engineering Heat Transfer*, D. Van Nostrand Company, Inc., New.York, 103, (1963).
4. Bayley, F. J., Owen, J. M. and Turner, A. B., *Heat Transfer*, Nelson (1972).

6

Forced convection: boundary layer principles

6.1 Introduction

In Chapter 1 it has been shown that to evaluate convection heat transfer, the magnitude of the coefficient h in Newton's equation has to be found. The study of convection centres round the behaviour of the fluid flowing past a surface, and the subject matter divides itself under various headings concerned with the type of flow situation or the method of analysis. This chapter shows how the convection

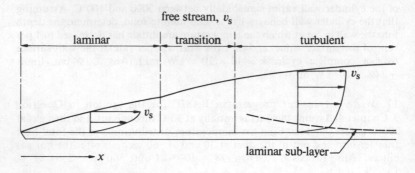

Fig. 6.1. Boundary layer growth on a flat plate.

coefficient may be determined by an approximate analytical method for simple cases of laminar flow. References to other methods will also be given.

Some familiarity with the flow of viscous fluids is assumed, and the growth of laminar boundary layers is illustrated in Figs. 6.1 and

6.2 where the turbulent boundary layer is also shown. Thus Fig. 6.1 shows the growth of laminar and turbulent boundary layers on a flat plate with a transition region occurring at

$$\frac{\rho v_s x}{\mu} \geqslant 5 \times 10^5 \tag{6.1}$$

where μ is the coefficient of molecular viscosity. The boundary layer exists as a result of the action of viscous shear within the fluid, the shear stress being proportional to the velocity gradient

$$\tau = \mu \frac{\mathrm{d}v_x}{\mathrm{d}y} \tag{6.2}$$

The group in equation (6.1), $\rho v_s x/\mu$, is the dimensionless *Reynolds number*, and is the ratio of momentum forces $\propto \rho v_s^2$, to shear forces $\propto \mu v_s/x$. Fig. 6.2 shows the growth of a laminar boundary layer in a tube

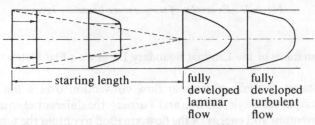

|—————— starting length ——————| fully developed laminar flow   fully developed turbulent flow

Fig. 6.2. *Boundary layer growth in a tube.*

with fully developed turbulent flow shown for comparison. The starting length is the length of tube required for the boundary layer to become fully developed. The velocity profiles follow closely the following equations:

for laminar flow: $\quad \dfrac{v}{v_a} = \dfrac{y}{r}\left(2 - \dfrac{y}{r}\right)$ $\tag{6.3}$

for turbulent flow: $\dfrac{v}{v_a} = \left(\dfrac{y}{r}\right)^{\frac{1}{4}}$ $\tag{6.4}$

where v is the velocity at distance y from the tube wall, v_a is the velocity at the axis.

Thermal boundary layers also exist. These are flow regions where the fluid temperature changes from the free stream value to the value at the surface. Examples in flow over a flat plate are shown in Fig. 6.3.

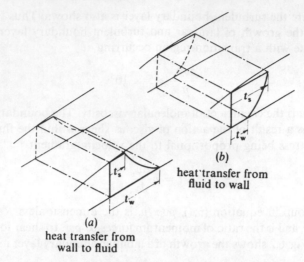

(b)

heat transfer from
fluid to wall

(a)

heat transfer from
wall to fluid

Fig. 6.3.   Examples of temperature boundary layers.

6.2 Equations of the Laminar Boundary Layer on a Flat Plate

In an exact analysis of laminar flow convection over a flat plate,
for example, see Bayley, Owen and Turner[1], the differential equations
of momentum and energy of the flow are used to obtain the tempéra-
ture gradient in the fluid at the wall, and hence the convection
coefficient. In this Section the laminar flow differential equations
will be derived together with integral equations for an approximate
analysis to be introduced in the next Section.

6.2.1 The Differential Equations of Continuity, Momentum and Energy

The control volume within the boundary layer in Fig. 6.4 is to be considered. For continuity, assuming steady state conditions with unit depth
and fluid desnity ρ, the mass flow rates in and out in the x-direction are

$$\rho v_x \mathrm{d}y$$

and

$$\rho \left(v_x + \frac{\partial v_x}{\partial x} \mathrm{d}x \right) \mathrm{d}y$$

respectively and hence the net flow into the element in the x-direction is

$$- \rho \frac{\partial v_x}{\partial x} \, dx \, dy$$

Similarly the net flow into the volume in the y-direction is

$$- \rho \frac{\partial v_y}{\partial y} \, dy \, dx$$

The total net flow in must be zero, hence

$$- \rho \left(\frac{\partial v_x}{\partial x} + \frac{\partial v_y}{\partial y} \right) dx \, dy = 0$$

Since ρ, dx and dy are not zero, it follows that

$$\frac{\partial v_x}{\partial x} + \frac{\partial v_y}{\partial y} = 0 \qquad (6.5)$$

The equation of momentum arises from the application of Newton's second law of motion to the element, assuming the fluid is

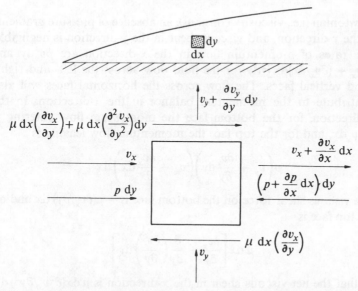

Fig. 6.4. *Element of boundary layer for continuity and momentum balance.*

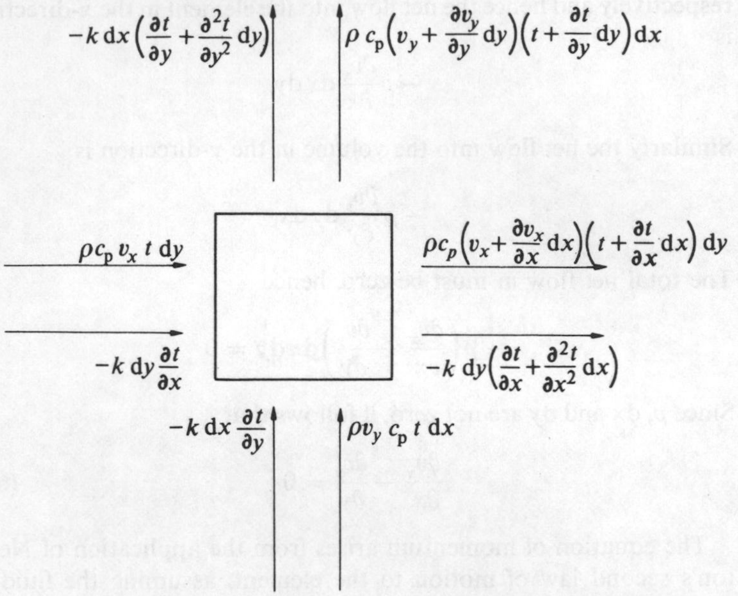

$$-k\,dx\left(\frac{\partial t}{\partial y}+\frac{\partial^2 t}{\partial y^2}\,dy\right) \qquad \rho\,c_p\left(v_y+\frac{\partial v_y}{\partial y}\,dy\right)\left(t+\frac{\partial t}{\partial y}\,dy\right)dx$$

$$\xrightarrow{\quad\rho c_p\,v_x\,t\,dy\quad} \qquad\qquad \xrightarrow{\quad\rho c_p\left(v_x+\frac{\partial v_x}{\partial x}dx\right)\left(t+\frac{\partial t}{\partial x}dx\right)dy\quad}$$

$$\xrightarrow{\quad-k\,dy\,\frac{\partial t}{\partial x}\quad} \qquad\qquad \xrightarrow{\quad-k\,dy\left(\frac{\partial t}{\partial x}+\frac{\partial^2 t}{\partial x^2}\,dx\right)\quad}$$

$$-k\,dx\,\frac{\partial t}{\partial y} \qquad\qquad \rho v_y\,c_p\,t\,dx$$

Fig. 6.5. *Element of boundary layer for energy balance.*

Newtonian (i.e., viscosity constant), an absence of pressure gradients in the y-direction, and viscous shear in the y-direction is negligible. The rates of momentum flow in the x-direction are $\rho v_x^2 dy$ and $\rho[v_x + (\partial v_x/\partial x)dx]^2 dy$ for the fluid flow across the left- and right-hand vertical faces. The flow across the horizontal faces will also contribute to the momentum balance in the x-direction. In the x-direction, for the bottom face the momentum flow entering is $\rho v_y v_x dx$, and for the top face the momentum flow leaving is

$$\rho\left(v_y + \frac{\partial v_y}{\partial y}dy\right)\left(v_x + \frac{\partial v_x}{\partial x}dx\right)dx$$

The viscous shear force on the bottom face is $-\mu(\partial v_x/\partial y)\,dx$ and on the top face is

$$\mu\,dx\left[\frac{\partial v_x}{\partial y} + \frac{\partial}{\partial y}\left(\frac{\partial v_x}{\partial y}\right)dy\right]$$

so that the net viscous shear in the x-direction is $\mu\,dx(\partial^2 v_x/\partial y^2)\,dy$. The pressure force on the left face is pdy, and on the right $-[p +$

$(\partial p/\partial x)dx]\,dy$ giving a net pressure force in the direction of motion of $-(\partial p/\partial x)\,dx\,dy$. Equating the sum of the net forces to the momentum flow out of the control volume in the x-direction gives, after neglecting second-order differentials and using the continuity equation:

$$\rho\left(v_x\frac{\partial v_x}{\partial x} + v_y\frac{\partial v_x}{\partial y}\right) = \mu\frac{\partial^2 v_x}{\partial y^2} - \frac{\partial p}{\partial x} \qquad (6.6)$$

The energy equation may now be deduced assuming constant properties and an absence of shear work as in a low velocity flow. Fig. 6.5 shows the energy terms involved, and it will be seen that there are four convective terms in addition to the conduction terms used in deriving equation (2.7). The energy balance is simply that rate of net conduction *in* + rate of net convection *in* = 0, hence

$$k\,dx\,dy\left(\frac{\partial^2 t}{\partial x^2} + \frac{\partial^2 t}{\partial y^2}\right) - \left[\rho c_p\left(v_x\frac{\partial t}{\partial x} + \frac{\partial v_x}{\partial x}t + \frac{\partial v_x}{\partial x}\frac{\partial t}{\partial x}dx\right)\right]dx\,dy$$

$$- \left[\rho c_p\left(v_y\frac{\partial t}{\partial y} + \frac{\partial v_y}{\partial y}t + \frac{\partial v_y}{\partial y}\frac{\partial t}{\partial y}dy\right)\right]dx\,dy = 0$$

Using the continuity equation and neglecting the second-order terms:

$$v_x\frac{\partial t}{\partial x} + v_y\frac{\partial t}{\partial y} = \alpha\left(\frac{\partial^2 t}{\partial x^2} + \frac{\partial^2 t}{\partial y^2}\right) \qquad (6.7)$$

The conduction in the x-direction is usually neglected in comparison with other terms and hence $\partial^2 t/\partial x^2$ may be dropped from equation (6.7). If in equation (6.7) the pressure gradient is assumed small and is neglected a similarity is then apparent between the equations of momentum and energy:

$$v_x\frac{\partial v_x}{\partial x} + v_y\frac{\partial v_x}{\partial y} = \nu\left(\frac{\partial^2 v_x}{\partial y^2}\right)$$

$$v_x\frac{\partial t}{\partial x} + v_y\frac{\partial t}{\partial y} = \alpha\left(\frac{\partial^2 t}{\partial y^2}\right)$$

ν is the kinematic viscosity or momentum diffusivity, μ/ρ, and $\nu/\alpha = (\mu/\rho)/(k/\rho c_p) = \mu c_p/k$, which is called the *Prandtl number, Pr*. If $\nu = \alpha$, then $Pr = 1$, and the pair of equations will lead to identical non-dimensionalised solutions of v_x and t as functions of y. The Prandtl number is the ratio of fluid properties controlling the velocity and temperature distributions, and it can vary between around 4×10^{-3} for a liquid metal to the order of 4×10^4 for a viscous oil.

6.2.2 The Integral Momentum and Energy Equations of the Laminar Boundary Layer

To consider the motion in the boundary layer, an elemental control volume is chosen that extends from the wall to just beyond the limit of the boundary layer in the y-direction, is dx thick in the x-direction, and has unit depth in the z-direction. This is shown in Fig. 6.6. An equation is sought which relates the net momentum outflow in the x-direction to the net force acting in the x-direction.

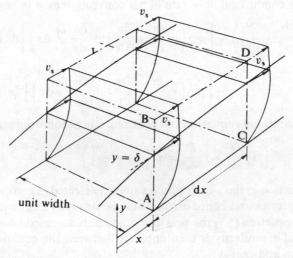

Fig. 6.6. Elemental control volume in laminar boundary layer.

The momentum flow across the face AB will be

$$\int_0^\delta \rho v_x^2 \, dy$$

Similarly, the momentum flow across the face CD will be

$$\int_0^\delta \rho v_x^2 \, dy + \frac{d}{dx} \int_0^\delta \rho v_x^2 \, dy \, dx$$

Fluid also enters the control volume across the face BD at the rate

$$\frac{d}{dx} \int_0^\delta \rho v_x \, dy \, dx$$

This is the difference between the fluid leaving across face CD and entering across face AB. The fluid entering across face BD has a velocity v_s in the x-direction, hence the flow of momentum into the control volume in the x-direction is

$$v_s \frac{d}{dx} \int_0^\delta \rho v_x \, dy \, dx$$

Hence the net outflow of momentum in the x-direction is

$$\frac{d}{dx} \int_0^\delta \rho v_x^2 \, dy \, dx - v_s \frac{d}{dx} \int_0^\delta \rho v_x \, dy \, dx$$

Pressure forces will act on faces AB and CD, and a shear force will act on face AC. There will be no shear force on face BD since this is at the limit of the boundary layer and $dv_x/dy = 0$. The net force acting on the control volume in the x-direction will be

$$p_x \delta - \left(p_x + \frac{dp_x}{dx} dx \right) \delta - \tau_w \, dx = - \delta \frac{dp_x}{dx} dx - \tau_w \, dx \qquad (6.8)$$

The pressure gradient may be neglected as small compared with the shear force at the wall, and the equality of the net momentum outflow to the net force gives

$$\frac{d}{dx} \int_0^\delta \rho v_x (v_s - v_x) \, dy = \tau_w \qquad (6.9)$$

This is the integral equation of motion in the laminar boundary layer, and was first derived by von Kármán.[2]

The integral energy equation may be derived in much the same way. In this case, a control volume extending beyond the limits of both temperature and velocity boundary layers may be considered initially, Fig. 6.7. The principle of conservation of energy applied to this control volume will involve the enthalpy and kinetic energy of fluid entering and leaving, and heat transfer by conduction at the wall. Kinetic energy may be neglected as being small in comparison

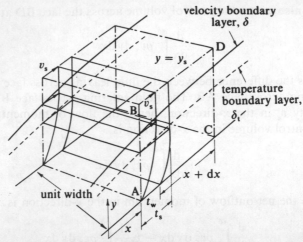

Fig. 6.7. Control volume in temperature and velocity boundary layers.

with other quantities. The enthalpy flow rate across face AB is

$$\int_0^{y_s} c_p \rho v_x t \, \mathrm{d}y$$

and across face CD

$$\int_0^{y_s} c_p \rho v_x t \, \mathrm{d}y + \frac{\mathrm{d}}{\mathrm{d}x} \int_0^{y_s} c_p \rho v_x t \, \mathrm{d}y \, \mathrm{d}x$$

Fluid will also enter the control volume across face BD at the rate

$$\frac{\mathrm{d}}{\mathrm{d}x} \int_0^{y_s} \rho v_x \, \mathrm{d}y \, \mathrm{d}x$$

Again this is the difference between the flow rate out at face CD and in at face AB. The enthalpy flow will be

$$c_p t_s \frac{\mathrm{d}}{\mathrm{d}x} \int_0^{y_s} \rho v_x \, \mathrm{d}y \, \mathrm{d}x$$

Finally, heat transfer by conduction across the wall at AC will amount to

$$-k \, \mathrm{d}x \left(\frac{\partial t}{\partial y} \right)_{y=0}$$

For conservation of energy:

$$c_p t_s \frac{d}{dx} \int_0^{y_s} \rho v_x \, dy \, dx - \frac{d}{dx} \int_0^{y_s} \rho c_p t v_x \, dy \, dx - k \, dx \left(\frac{\partial t}{\partial y} \right)_{y=0} = 0 \quad (6.10)$$

Beyond the limit of the temperature boundary layer, the temperature is constant at t_s, and hence the integration need only be taken up to $y = \delta_t$. Equation (6.10) therefore gives

$$\frac{d}{dx} \int_0^{\delta_t} (t_s - t) v_x \, dy - \alpha \left(\frac{\partial t}{\partial y} \right)_{y=0} = 0 \quad (6.11)$$

This, then, is the integral energy equation of the laminar boundary layer.

6.3 Laminar Forced Convection on a Flat Plate

The integral equations (6.9) and (6.11) will now be applied to the problem of laminar forced convection on a flat plate. The method is due to Eckert.[3] The analysis assumes the viscosity is uniform with temperature. The first step is to use the integral equation of motion to derive an equation for boundary layer thickness. The velocity contour may, for example, be assumed a polynomial

$$v_x = a + by + cy^2 + dy^3$$

where a, b, c, and d are constants. The constants may be found by applying known boundary conditions. Thus $v_x = 0$ at $y = 0$, and hence $a = 0$. Also $v_x = v_s$ at $y = \delta$, and $(\partial v_x / \partial y)_\delta = 0$ at $y = \delta$. Further, since both v_x and v_y are zero at $y = 0$, it follows from (6.5) that $\partial^2 v_x / \partial y^2 = 0$, at $y = 0$. These results lead to

$$b = \frac{3}{2} \frac{v_s}{\delta}, \quad c = 0, \quad d = -\frac{v_s}{2\delta^3}$$

and hence

$$\frac{v_x}{v_s} = \frac{3}{2} \left(\frac{y}{\delta} \right) - \frac{1}{2} \left(\frac{y}{\delta} \right)^3 \quad (6.12)$$

Applying the integral equation of motion,

$$\frac{d}{dx} \int_0^{\delta} \rho v_x (v_s - v_x) \, dy = \tau_w$$

$$= \frac{d}{dx} \int_0^{\delta} \rho v_s^2 \left[\frac{3}{2} \left(\frac{y}{\delta} \right) - \frac{1}{2} \left(\frac{y}{\delta} \right)^3 \right] \cdot \left[1 - \frac{3}{2} \left(\frac{y}{\delta} \right) + \frac{1}{2} \left(\frac{y}{\delta} \right)^3 \right] dy$$

$$= \mu \left(\frac{dv_x}{dy} \right)_{y=0}$$

The wall shear stress is found by considering the velocity gradient at $y = 0$; this is found to be $3v_s/2\delta$. The above equation leads to

$$\frac{d}{dx} \rho v_s^2 \frac{39\delta}{280} = \frac{3}{2}\mu \frac{v_s}{\delta}$$

$$\therefore \quad \rho v_s^2 \, d\delta = \frac{3}{2} \cdot \frac{280}{39} \frac{\mu v_s}{\delta} \, dx$$

$$\therefore \quad \delta \, d\delta = \frac{140}{13} \cdot \frac{v}{v_s} \, dx$$

On integration

$$\frac{\delta^2}{2} = \frac{140vx}{13v_s} + C$$

$C = 0$, since $\delta = 0$ at $x = 0$

$$\therefore \quad \delta^2 = \frac{280vx}{13v_s}$$

or

$$\frac{\delta}{x} = \frac{4\cdot64}{Re_x^{\frac{1}{2}}} \tag{6.13}$$

This result, due to Pohlhausen,[4] is required later on in considering the integral energy equation.

The temperature distribution in the boundary layer is assumed to follow a similar law to the velocity distribution. Thus:

$$\theta = (t - t_w) = dy + ey^2 + fy^3$$

where, again, d, e, and f are constants. The boundary conditions are that at $y = \delta_t$ (the thickness of the temperature boundary layer), $\theta = \theta_s$ and also $(\partial\theta/\partial y)_{\delta_t} = 0$. Also, from equation (6.7) it follows that $(\partial^2\theta/\partial y^2)_{y=0} = 0$ because v_x and v_y are both zero at $y = 0$. From these conditions it follows that

$$d = \frac{3}{2}\frac{\theta_s}{\delta_t}, \quad e = 0, \quad f = -\frac{\theta_s}{2\delta_t^3}$$

and hence

$$\frac{\theta}{\theta_s} = \frac{3}{2}\left(\frac{y}{\delta_t}\right) - \frac{1}{2}\left(\frac{y}{\delta_t}\right)^3 \tag{6.14}$$

Turning to the integral energy equation, the substitutions
$\theta = (t - t_w)$ and $\theta_s = (t_s - t_w)$ are made to give

$$\frac{d}{dx} \int_0^{\delta_t} (\theta_s - \theta)v_x \, dy - \alpha \left(\frac{\partial \theta}{\partial y}\right)_{y=0} = 0 \qquad (6.15)$$

From the temperature equation (6.14) it follows that

$$\alpha \left(\frac{\partial \theta}{\partial y}\right)_{y=0} = \alpha \frac{3\theta_s}{2\delta_t}$$

This result is substituted in equation (6.15) together with the expressions for θ and v_x to give:

$$\frac{d}{dx} \int_0^{\delta_t} \left[\theta_s - \frac{3}{2}\left(\frac{y}{\delta_t}\right)\theta_s + \frac{1}{2}\left(\frac{y}{\delta_t}\right)^3\theta_s\right] \cdot \left[\frac{3}{2}\left(\frac{y}{\delta}\right)v_s - \frac{1}{2}\left(\frac{y}{\delta}\right)^3 v_s\right] dy = \alpha \frac{3\theta_s}{2\delta_t}$$

A useful substitution is that $\lambda = \delta_t/\delta$.

$$\therefore \quad \frac{d}{dx} \theta_s v_s \int_0^{\delta_t} \left[1 - \frac{3}{2}\left(\frac{y}{\lambda\delta}\right) + \frac{1}{2}\left(\frac{y}{\lambda\delta}\right)^3\right] \cdot \left[\frac{3}{2}\left(\frac{y}{\delta}\right) - \frac{1}{2}\left(\frac{y}{\delta}\right)^3\right] dy$$

$$= \alpha \frac{3\theta_s}{2\lambda\delta}$$

This then leads to

$$\frac{d}{dx}\left[\theta_s v_s \delta_t \left(\frac{3\lambda}{20} - \frac{3\lambda^3}{280}\right)\right] = \alpha \frac{3\theta_s}{2\lambda\delta}$$

It is convenient here to neglect the term $3\lambda^3/280$ as being small in comparison with $3\lambda/20$. This is justified since λ has the value of 1 if $Pr = 1$, and will not be far removed from 1 at other values of Pr fairly close to 1. Hence

$$\frac{d}{dx}\frac{3\lambda\delta_t}{20} = \alpha \frac{3}{2\lambda\delta v_s}$$

$$\therefore \quad \frac{d}{dx}(\lambda^2\delta) = \frac{10\alpha}{v_s\lambda\delta}$$

$$\therefore \quad 2\lambda\delta\frac{d\lambda}{dx} + \lambda^2\frac{d\delta}{dx} = \frac{10\alpha}{v_s\lambda\delta}$$

$$\therefore \quad 2\lambda^2\delta^2\frac{d\lambda}{dx} + \lambda^3\delta\frac{d\delta}{dx} = \frac{10\alpha}{v_s}$$

Equation (6.13) for δ may now be substituted.

$$\frac{\delta}{x} = \frac{4\cdot64}{Re_x^{\frac{1}{2}}}, \quad \text{and hence} \quad \delta = 4\cdot64\left(\frac{xv}{v_s}\right)^{\frac{1}{2}}$$

$$\therefore \quad 2\lambda^2\frac{21\cdot6xv}{v_s}\cdot\frac{d\lambda}{dx} + \frac{\lambda^3}{2}\cdot\frac{21\cdot6v}{v_s} = \frac{10\alpha}{v_s}$$

$$\therefore \quad 4\lambda^2x\frac{d\lambda}{dx} + \lambda^3 = \frac{0\cdot93\alpha}{v} \tag{6.16}$$

This equation may be solved by making the substitution $\lambda^3 = p$, and $p = x^n$, and the solution obtained is:

$$\left(\frac{\delta_t}{\delta}\right)^3 = \frac{0\cdot93}{Pr} + \frac{M}{x^{\frac{3}{4}}}$$

noting that $\alpha/v = Pr$, and M is a constant of integration. The thickness of the thermal boundary layer will be 0 at the beginning of the heated section, at $x = x_h$, say, and hence

$$M = -\frac{0\cdot93x_h^{\frac{3}{4}}}{Pr}$$

and finally:

$$\left(\frac{\delta_t}{\delta}\right)^3 = \frac{0\cdot93}{Pr}\left[1 - \left(\frac{x_h}{x}\right)^{\frac{3}{4}}\right]$$

This result may be simplified further by assuming that the plate is heated along its entire length, or $x_h = 0$, in Fig. 6.8,

$$\left.\begin{array}{l}\text{hence} \quad \dfrac{\delta_t}{\delta} = \left(\dfrac{0\cdot93}{Pr}\right)^{\frac{1}{3}} \\[3mm] \text{or, approximately,} \quad \dfrac{\delta_t}{\delta} = \dfrac{1}{Pr^{\frac{1}{3}}}\end{array}\right\} \tag{6.17}$$

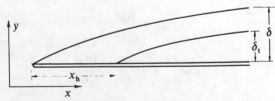

Fig. 6.8. *Laminar forced convection on a flat plate. Heating commences x_h from the leading edge.*

Using the equations for δ_t, δ, and the value of $(\partial\theta/\partial y)_{y=0}$ it is possible to determine the heat transfer at the wall, since

$$q_w = k\left(\frac{\partial\theta}{\partial y}\right)_{y=0} = k\frac{3\theta_s}{2\delta_t}, \quad \text{from (6.14)}$$

The heat transfer rate at the wall is expressed non-dimensionally. q_w/θ_s is the heat transfer coefficient h, and the group hx/k is the dimensionless Nusselt number, Nu. It is interpreted as the ratio of two lengths, the characteristic linear dimension of the system, and an equivalent conducting film of thickness δ_t'. Figure 6.9 shows

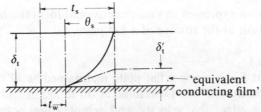

Fig. 6.9. *To illustrate the significance of the Nusselt number.*

how δ_t' is defined. The heat transfer at the wall is $q_w = h\theta_s$ and may be expressed as $q_w = (k/\delta_t')\theta_s$. It follows that $h = k/\delta_t'$ and hence

$$Nu = \frac{hx}{k} = \frac{x}{\delta_t'}$$

The linear dimension of the system is generally large in comparison with δ_t'.

A Nusselt number may therefore be obtained:

$$Nu_x = \frac{q_w x}{\theta_s k} = \frac{3x}{2\delta_t} = \frac{3x\, Re_x^{\frac{1}{2}}\, Pr^{\frac{1}{3}}}{2(0\cdot93)^{\frac{1}{3}}4\cdot64x}$$

using (6.17) to eliminate δ_t, and (6.13) to eliminate δ.

$$\therefore \quad Nu_x = 0\cdot332 Re_x^{\frac{1}{2}}\, Pr^{\frac{1}{3}} \tag{6.18}$$

This gives the local Nusselt number at some distance x from the leading edge of the plate. The average value of the convection coefficient h, over the distance 0 to x is given by:

$$\bar{h} = \frac{1}{x}\int_0^x h\,dx$$

where

$$h = 0.332k \left(\frac{v_s}{vx}\right)^{\frac{1}{2}} Pr^{\frac{1}{3}}, \quad \text{from equation (6.18)}$$

$$\therefore \quad \bar{h} = \frac{k}{x} \frac{0.332}{\frac{1}{2}} \left(\frac{v_s x}{v}\right)^{\frac{1}{2}} Pr^{\frac{1}{3}}$$

$$= 0.664k \left(\frac{v_s}{vx}\right)^{\frac{1}{2}} Pr^{\frac{1}{3}}$$

and

$$\overline{Nu_x} = 0.664 \, Re_x^{\frac{1}{2}} \, Pr^{\frac{1}{3}} \tag{6.19}$$

This equation expresses in non-dimensional form the heat transfer by convection at the surface of a flat plate.

EXAMPLE 6.1

Air flows at 5 m/s along a flat plate maintained at 77°C. The bulk air temperature is 27°C. Determine at 0·1 m ,0·5 m and 1·0 m from the leading edge, the velocity and temperature boundary layer thicknesses, and the local and average convection coefficients. Use mean properties of air from Table A6.

Solution. At 325 K, $\rho = 1.087 \, \text{kg/m}^3$, $k = 28.1 \times 10^{-6} \, \text{kW/(m K)}$, $\mu = 1.965 \times 10^{-5}$ Pa s, and $Pr = 0.703$.

The Reynolds numbers at $x = 0.1, 0.5$ and 1.0 m with $v_s = 5.0$ m/s, together with the boundary layer thicknesses using $\delta/x = 4.64/Re_x^{\frac{1}{2}}$, equation (6.13), and $\delta_t/\delta = 1/Pr^{\frac{1}{3}}$, equation (6.17), and the local and average coefficients using equations (6.18) and (6.19), are calculated and tabulated below:

	Re_x	δ mm	δ_t mm	h kW/(m² K)	$\bar{h}$ kW/(m² K)
$x = 0.1$	2.76×10^4	2·79	3·14	13.8×10^{-3}	27.6×10^{-3}
0·5	1.38×10^5	6·26	7·04	6.15×10^{-3}	12.3×10^{-3}
1·0	2.76×10^5	8·92	10·02	4.37×10^{-3}	8.74×10^{-3}

6.4 Laminar Forced Convection in a Tube

Laminar forced convection in a tube will be considered for the case of fully developed flow and constant heat flux at the wall. For fully developed flow it may be assumed that the velocity profile has a parabolic shape. It is first necessary to derive the energy equation for flow in a tube. To do this, a small cylindrical element of flow may be considered, as in Fig. 6.10. The element is of length dx, radius r on the inside, and radius $r + dr$ on the outside. Energy will flow into and out of the element in the radial direction by conduction,

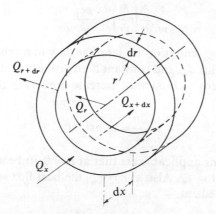

Fig. 6.10. Element of laminar flow in a tube.

and in the axial direction by convection. Conduction into the element is

$$Q_r = -k2\pi r \, dx \frac{\partial t}{\partial r}$$

Over the change of radius dr, this conduction rate will change by

$$\frac{\partial Q_r}{\partial r} dr = -k2\pi \, dx \frac{\partial}{\partial r}\left(r\frac{\partial t}{\partial r}\right) dr \qquad (6.20)$$

This change in conduction rate is accounted for by the difference between the convection rates into and out of the element in the axial direction. The axial velocity through the element is constant but the temperature changes in the axial direction. The rate of convection into the element is

$$2\pi r \, dr \, \rho v c_p t$$

and out of the element, it is

$$2\pi r \, dr \, \rho v c_p \left(t + \frac{\partial t}{\partial x} dx\right)$$

Hence, the difference is

$$2\pi r\, dr\, \rho v c_p \frac{\partial t}{\partial x} dx$$

The sum of this and the right-hand side of (6.20) is zero, hence on cancelling terms,

$$\frac{1}{vr}\frac{\partial}{\partial r}\left(r\frac{\partial t}{\partial r}\right) = \frac{\rho c_p}{k}\frac{\partial t}{\partial x} \tag{6.21}$$

This is the energy equation for laminar flow in a tube. With a constant wall heat flux q_w, and constant fluid properties, the temperature of the fluid (at any radius) must increase linearly in the direction of flow, so that

$$\frac{\partial t}{\partial x} = \text{constant}$$

Other conditions applicable are that at $r = 0$, (tube axis), $\partial t/\partial r = 0$ and at $r = r_w$, $t = t_w$. Also at $r = r_w$, the heat flux is related to the temperature gradient,

$$q_w = -k\left(\frac{\partial t}{\partial r}\right)_{r_w}$$

Since $\partial t/\partial x$ is assumed constant, equation (6.21) reduces to a total differential equation. The velocity v is a function of the velocity at the axis of the tube, v_a, and the radius r. The assumed parabolic velocity distribution, equation (6.3), expressed in terms of r measured from the axis is

$$\frac{v}{v_a} = 1 - \left(\frac{r}{r_w}\right)^2$$

where r_w is the wall radius.

This result must be substituted into equation (6.21) before integrating. Hence, after re-arrangement,

$$\frac{\partial}{\partial r}\left(r\frac{\partial t}{\partial r}\right) = \frac{1}{\alpha}\frac{\partial t}{\partial x}v_a\left[1 - \left(\frac{r}{r_w}\right)^2\right]r$$

This is integrated to give

$$r\frac{\partial t}{\partial r} = \frac{1}{\alpha}\frac{\partial t}{\partial x}v_a\left(\frac{r^2}{2} - \frac{r^4}{4r_w^2}\right) + C_1$$

and after a second integration, gives

$$t = \frac{1}{\alpha}\frac{\partial t}{\partial x}v_a\left(\frac{r^2}{4} - \frac{r^4}{16r_w^2}\right) + C_1 \ln r + C_2 \qquad (6.22)$$

C_1 and C_2 are constants of integration to be found from boundary conditions. Since $\partial t/\partial r = 0$ at $r = 0$, it follows that $C_1 = 0$. The other boundary condition is that at $r = r_w$, $t = t_w$. Hence,

$$t_w = \frac{1}{\alpha}\frac{\partial t}{\partial x}v_a\left[\frac{r_w^2}{4} - \frac{r_w^2}{16}\right] + C_2$$

$$\therefore \quad C_2 = t_w - \frac{1}{\alpha}\frac{\partial t}{\partial x}v_a\frac{3r_w^2}{16}$$

Hence equation (6.22) becomes, after some re-arrangement,

$$t = \frac{1}{\alpha}\frac{\partial t}{\partial x}v_a r_w^2\left[\frac{1}{4}\left(\frac{r}{r_w}\right)^2 - \frac{1}{16}\left(\frac{r}{r_w}\right)^4 - \frac{3}{16}\right] + t_w \qquad (6.23)$$

This equation may be expressed as a temperature difference, $\theta = t - t_w$. Further, if θ_a is the temperature difference between the axis, where $r = 0$, and the wall, θ_a may be found from equation (6.23) by putting $r = 0$. Hence,

$$\theta_a = \frac{1}{\alpha}\frac{\partial t}{\partial x}v_a r_w^2\left(-\frac{3}{16}\right) \qquad (6.24)$$

The temperature profile may be expressed non-dimensionally by dividing equation (6.23) by equation (6.24).

$$\therefore \quad \frac{\theta}{\theta_a} = 1 - \frac{4}{3}\left(\frac{r}{r_w}\right)^2 + \frac{1}{3}\left(\frac{r}{r_w}\right)^4 \qquad (6.25)$$

The equation for heat transfer at the wall may be obtained by considering the temperature gradient at $r = r_w$. Thus, from equation (6.25)

$$\left(\frac{d\theta}{dr}\right)_{r_w} = \theta_a\left(-\frac{8}{3r_w} + \frac{4}{3r_w}\right) = -\frac{4\theta_a}{3r_w}$$

and

$$q_w = - k\left(\frac{d\theta}{dr}\right)_{r_w} = \frac{4k\theta_a}{3r_w} = h\theta_a$$

$$\therefore \quad h = \frac{4k}{3r_w}$$

In terms of the Nusselt number, Nu_d

$$Nu_d = \frac{hd}{k} = \frac{4k}{3r_w}\frac{2r_w}{k} = \frac{8}{3} \qquad (6.26)$$

This value of Nusselt number is based on the difference in temperature between the tube axis and the wall. However, from a practical point of view it is more convenient to consider the difference in temperature between the bulk value and the wall. The bulk temperature is the mean temperature of the fluid, and the temperature difference required is given by

$$\theta_m = \frac{\int_0^{r_w} 2\pi r \, dr \rho v c_p \theta}{\int_0^{r_w} 2\pi r \, dr \rho v c_p}$$

Introducing equations for v and θ, this becomes

$$\theta_m = \frac{\int_0^{r_w} 2\pi \rho c_p v_a \theta_a \left[1 - \left(\frac{r}{r_w}\right)^2\right]\left[1 - \frac{4}{3}\left(\frac{r}{r_w}\right)^2 + \frac{1}{3}\left(\frac{r}{r_w}\right)^4\right] r \, dr}{\int_0^{r_w} 2\pi \rho c_p v_a \left[1 - \left(\frac{r}{r_w}\right)^2\right] r \, dr}$$

This, on integration, gives $\theta_m = \frac{44}{72}\theta_a$.
The heat transfer at the wall is,

$$q_w = - k\left(\frac{d\theta}{dr}\right)_{r_w} = \frac{4k\theta_a}{3r_w} = \frac{4k}{3r_w}\cdot\frac{72}{44}\,\theta_m$$

This is now equivalent to

$$q_w = h\theta_m$$

Hence

$$h = \frac{4k}{3r_w} \cdot \frac{72}{44}$$

and

$$Nu_d = \frac{hd}{k} = \frac{4k}{3r_w} \cdot \frac{72}{44} \cdot \frac{2r_w}{k}$$

$$= \frac{8}{3} \cdot \frac{72}{44}$$

$$= 4 \cdot 36 \qquad (6.27)$$

These results are independent of Reynolds number because, for fully developed flow, the boundary layer thickness is equal to the tube radius.

Example 6.2

Water at a mean temperature of 40°C flows at a mean velocity of 0·1 m/s in a 3 mm bore tube having a constant wall flux of 1·0 kW/m². Determine the temperature of the water as a function of radius, using equation (6.23). Use fluid properties from Table A5.

Solution. It is necessary to obtain $\partial t/\partial x$, also v_a the velocity of water at the centre of the tube.

Consider an energy balance on 1 m of tube:

$$q_w \pi d \times 1 = \text{mass flow} \times c_p \times \text{temperature rise}$$

$$\therefore \frac{\partial t}{\partial x} = q_w \pi d/(\text{mass flow} \times c_p)$$

$$= \frac{1 \cdot 0 \times \pi \times 3}{4 \cdot 178 \times 1000} \times \frac{4 \times 10^6}{\pi \times 3^2 \times 0 \cdot 1 \times 994 \cdot 6}$$

where $c_p = 4 \cdot 178$ kJ/(kg K), and $\rho = 994 \cdot 6$ kg/m³

$$\therefore \partial t/\partial x = 3 \cdot 22 \, \text{K/m}$$

The laminar flow velocity distribution is given by

$$(v/v_a) = 1 - (r/r_w)^2$$

where v_a = velocity at $r = 0$, hence for continuity:

$$v_m \pi r_w^2 = \int_0^{r_w} 2\pi r[v_a - v_a(r/r_w)^2]\,dr$$

$$= \pi r_w^2 v_a - \frac{2\pi r_w^2 v_a}{4} = \pi r_w^2 v_a/2$$

$$\therefore v_a = 2v_m$$

With $\alpha = 15.1 \times 10^{-8}$, equation (6.23) becomes:

$$t = \frac{10^8}{15.1} \times 3.22 \times 0.1 \times 2 \times \frac{1.5^2}{10^6}\left[\frac{1}{4}\left(\frac{r}{r_w}\right)^2 - \frac{1}{16}\left(\frac{r}{r_w}\right)^4 - \frac{3}{16}\right] + t_w$$

$$= 9.6\left[\frac{1}{4}\left(\frac{r}{r_w}\right)^2 - \frac{1}{16}\left(\frac{r}{r_w}\right)^4 - \frac{3}{16}\right] + t_w$$

At $r = r_w$, $t = t_w$; at $r = 0$, $t = t_w - 1.8°C$.

PROBLEMS

1. Derive the heat flow equation of the boundary layer

$$\frac{d}{dx}\int_0^l (\theta_s - \theta)U\,dy = \alpha\left(\frac{d\theta}{dy}\right)_0$$

and apply this equation to 'slug' flow of a liquid metal along a plate of uniform temperature to find the thickness of the temperature boundary layer. θ would be the liquid metal temperature relative to the plate temperature. Assume that the temperature profile in the boundary layer can be described by an equation of the form

$$\theta = a \sin\left[b\left(\frac{y}{\delta_t}\right) + c\right]$$

where a, b, and c are constants to be determined from the boundary conditions.

Hence prove that the local Nusselt number N_x is given by

$$N_x = \sqrt{\left(\frac{\pi - 2}{4}\right)}\sqrt{(R_x P)} = 0.534\sqrt{(R_x P)}$$

It can be shown that if the velocity profile can be approximated by an equation of the form

$$U = d \sin\left[e\left(\frac{y}{\delta}\right) + f\right]$$

the velocity boundary layer thickness is then given by

$$\frac{\delta}{x} = \sqrt{\left[\frac{2\pi^2}{(4 - \pi)R_x}\right]}$$

Show that for a liquid metal of $P = 0.01$ the temperature boundary layer thickness is approximately equal to 6δ. (*University of Bristol*).

2. Prove that, in hydrodynamically fully-developed laminar flow through a tube, the temperature field is determined by the following partial differential equation

$$\frac{1}{Ur}\frac{\partial}{\partial r}\left(r\frac{\partial t}{\partial r}\right) = \frac{1}{\alpha}\left(\frac{\partial t}{\partial x}\right)$$

where r is the distance from the axis of the tube, and U is the velocity at r.

Hence derive an equation for the fully developed temperature profile, when the heat flux q_w is constant along the wall of the tube. You may assume that the velocity profile is given by

$$\frac{U}{U_0} = 1 - \left(\frac{r}{R}\right)^2$$

Show that the temperature profile can be put into dimensionless form as

$$\frac{t - t_w}{t_0 - t_w} = \frac{\theta}{\theta_0} = 1 - \frac{4}{3}\left(\frac{r}{R}\right)^2 + \frac{1}{3}\left(\frac{r}{R}\right)^4$$

where t, t_0, and t_w are the local, axial, and wall temperatures respectively, and R is the radius of the tube. Also show that the Nusselt number

$$\frac{q_w d}{\theta_0 k} = \frac{8}{3}$$

Explain, by writing down the initial equations, how you would derive the Nusselt number $q_w d/\theta_m k$, where θ_m is the bulk temperature of the fluid relative to the wall. (*University of Bristol*).

3. Show that if a flat plate has a heated section commencing at x_h from the leading edge, the local Nusselt number at distance x from the leading edge, $(x > x_h)$, is given by:

$$Nu_x = 0.332\,Re_x^{\frac{1}{2}}\,Pr^{\frac{1}{3}}(1 - (x_h/x)^{\frac{3}{4}})^{-\frac{1}{3}}$$

Determine the velocity and thermal boundary layer thicknesses and the local heat transfer rate at 1 m from the leading edge of a plate heated 0·5 m from the leading edge, for air at 27°C flowing over the plate at 0·5 m/s, if the temperature of the heated section is 127°C. (Ans. $\delta = 0.0298$ m, $\delta_t = 0.0243$ m, 0.184 kW/m^2.)

4. The velocity in the boundary layer of a stream of air flowing over a flat plate can be represented by

$$\frac{u}{U} = \frac{3}{2}\left(\frac{y}{\delta}\right) - \frac{1}{2}\left(\frac{y}{\delta}\right)^3$$

where U is the main stream velocity, u the velocity at a distance y from the flat plate within the boundary layer of thickness δ. The variation of boundary layer thickness along the plate may be taken as

$$\delta/x = 4\cdot64(Re_x)^{-\frac{1}{2}}$$

If the plate is heated to maintain its surface at constant temperature show that the average Nusselt number over a distance x from the leading edge of the hot plate is

$$Nu = 0\cdot66(Pr)^{\frac{1}{3}}(Re_x)^{\frac{1}{2}}$$

(*University of Leeds*).

5. If in laminar flow heat transfer on a flat plate the velocity distribution is given by $V_x = V_s(y/\delta)$, and assuming in this case that there is no shear at the limit of the boundary layer, show that the boundary layer thickness is given by

$$\delta/x = 3\cdot46/Re_x^{\frac{1}{2}}$$

where δ is the boundary layer thickness at x from the leading edge. Also show that the average Nusselt number at x is given by

$$\overline{Nu_x} = 0\cdot73\,Re_x^{\frac{1}{2}}\,Pr^{\frac{1}{3}}$$

with heating commencing at $x = 0$.

REFERENCES

1. Bayley, F. J., Owen, J. M. and Turner, A. B. *Heat Transfer*, Nelson (1972).
2. Kármán, T. von, Z. *angew. Math. u. Mech.*, Vol. 1, 233 (1921).
3. Eckert, E. R. G. and Drake, R. M. *Analysis of Heat and Mass Transfer*, McGraw-Hill, New York (1972).
4. Pohlhausen, K. Z. *angew. Math. u. Mech.*, Vol. 1, 252 (1921).

7

Forced convection: Reynolds analogy and dimensional analysis

Consideration of convection has so far been limited to laminar flow. For turbulent flow, it is possible to introduce additional terms into the momentum and energy equations to account for the presence of turbulence, and to obtain numerical solutions to the finite difference forms of the equations.[1,2] However, these methods have only become possible with the use of the more recent and more powerful generations of digital computer, and at an introductory level the more classical approaches will be followed.

7.1 Reynolds Analogy

The approach to forced convection known as Reynolds analogy is based on similarities between the equations for heat transfer and shear stress, or momentum transfer. The original ideas were due to Reynolds[3,4] and the analogy has been subsequently modified and extended by others.

The equation for shear stress in laminar flow, (6.4), may be written as

$$\tau = \rho v \frac{dv}{dy} \tag{7.1}$$

where v is the kinematic viscosity, μ/ρ. A similar equation may be written for shear stress in turbulent flow. A term ε, eddy diffusivity, is introduced, which enables the shear stress due to random turbulent motion to be written

$$\tau_t = \rho \varepsilon \frac{dv}{dy} \tag{7.2}$$

101

When turbulent flow exists, the viscous shear stress is also present which may be added to τ_t. The total shear stress in turbulent flow is thus

$$\tau = \rho(v + \varepsilon)\frac{dv}{dy} \tag{7.3}$$

ε is not a property of the fluid as v is. It depends on several factors such as the Reynolds number of the flow and the turbulence level. Its value is generally many times greater than v.

7.1.1 Shear Stress at the Solid Surface

In developing Reynolds analogy the heat transfer at the surface of a flat plate or of a tube is ultimately compared with the shear stress acting at that surface. This shear stress is obtained by substituting $(dv/dy)_{y=0}$ into the equation for τ. Thus, for laminar flow on a flat plate, x from the leading edge, with the Reynolds number Re_x based on the free stream velocity and x,

$$Cf = \frac{0.647}{Re_x^{\frac{1}{2}}} \tag{7.4}$$

where Cf is the skin friction coefficient defined as $\tau_w/\frac{1}{2}\rho v_s^2$. v_s is the free stream velocity. An average value Cd for the length x is found to be $2Cf$ for laminar flow, where Cf is the local value at x. The derivative of the turbulent velocity profile substituted into (7.2) leads to an infinite shear stress at the wall. This is overcome by assuming the existence of a laminar sub-layer, as in Fig. 6.1. For turbulent flow on a flat plate, Cf and Cd are given by

$$Cf = 0.0583(Re_x)^{-\frac{1}{5}} \tag{7.5}$$

and

$$Cd = \frac{0.455}{(\log Re_x)^{2.58}} \tag{7.6}$$

Equation (7.6) is an empirical relationship,[5] which takes into account the laminar and turbulent portions of the boundary layer.

The ratio of the velocity at the limit of the laminar sublayer to the free stream velocity is also of importance, as will be seen later; this is a function of the Reynolds number at x from the leading edge:

$$\frac{v_b}{v_s} = \frac{2.12}{(Re_x)^{0.1}} \tag{7.7}$$

Corresponding relationships for flow in tubes are usually expressed in terms of a friction factor f, which is four times larger than Cf in terms of the surface shear stress. Thus $f = 4\tau_w/\frac{1}{2}\rho v_m^2$, where v_m is the mean velocity of flow

$$\text{In laminar flow,} \quad f = \frac{64}{Re_d} \quad (7.8)$$

$$\text{and in turbulent flow,} \quad f = \frac{0\cdot308}{(Re_d)^{\frac{1}{4}}} \quad (7.9)$$

and

$$\frac{v_b}{v_m} = \frac{2\cdot44}{(Re_d)^{\frac{1}{8}}} \quad (7.10)$$

The derivations of these relationships may be found in the more advanced texts on heat transfer, or fluid mechanics, e.g., refs. 6, 14.

The friction factors quoted above are for smooth surfaces. Values are increased if the surface is rough. For any *tube* surface, the average wall shear stress τ_w acting over a length L can be found by considering the forces acting. Thus, if Δp is the pressure loss and d the tube diameter, the pressure force $\Delta p \pi d^2/4$ is equal to the wall shear force $\tau_w \pi dL$, assuming the tube is horizontal.

7.1.2 Heat Transfer across the Boundary Layer

Equations for heat transfer across the boundary layer are written in analogous form to (7.1) and (7.3). Thus in laminar flow, heat transfer *across* the flow can only be by conduction, so Fourier's law may be written as

$$q = -\rho c_p \alpha \frac{dt}{dy} \quad (7.11)$$

In turbulent conditions energy will also be carried across the flow by random turbulent motion, and the heat flux may be written

$$q = -\rho c_p (\alpha + \varepsilon_q) \frac{dt}{dy} \quad (7.12)$$

where ε_q is the thermal eddy diffusivity, a term analogous to ε. The basis of Reynolds analogy is to compare equations (7.1) and (7.11) for laminar flow, and equations (7.3) and (7.12) for turbulent flow.

In equations (7.3) and (7.12) it has been seen that the ratio v/α is the Prandtl number; similarly $\varepsilon/\varepsilon_q$ is known as the turbulent Prandtl number, though this is not a property of the fluid as is v/α.

Some initial assumptions must now be made. The first is that $\varepsilon = \varepsilon_q$. This means that if an eddy of fluid, at a certain temperature and possessing a certain velocity, is transferred to a region at a different state, then it assumes its new temperature and velocity in equal times. This assumption is found by experiment to be approximately true. ($\varepsilon_q/\varepsilon$ varies between 1 and 1·6. For a review of this subject, see ref. 6.) A second assumption is that q and τ have the same ratio at all values of y. This will be true when velocity and temperature profiles are identical. Identical profiles occur in laminar flow when the Prandtl number of the fluid is 1. In turbulent flow, with $\varepsilon = \varepsilon_q$, the groups responsible for velocity and temperature distributions, $(v + \varepsilon)$ and $(\alpha + \varepsilon_q)$, are also equal when $Pr = 1$. Further, even when the Prandtl number is not 1, $(v + \varepsilon)$ and $(\alpha + \varepsilon_q)$ will be nearly equal, since ε and ε_q are very much greater than v and α.

The simple Reynolds analogy is valid when $Pr = 1$, and the Prandtl–Taylor modification[7, 8] which takes into account a varying Pr is valid for a fairly restricted range, say $0·5 < Pr < 2·0$.

7.1.3 The Simple Reynolds Analogy

With the assumptions noted above it is now possible to proceed to a consideration of the simple analogy. Flow is assumed to be either all laminar or all turbulent, and $Pr = 1$. By comparing equations (7.1) and (7.11) for laminar flow, it follows that

$$\frac{q}{\tau} = -\frac{k}{\mu}\frac{dt}{dv} \tag{7.13}$$

This gives the ratio of q/τ at some arbitrary plane in the flow. Noting that q/τ has the same value anywhere in the y-direction, it is possible to express q_w/τ_w at the wall in terms of free stream and wall temperatures and velocities.
Thus

$$\frac{q_w}{\tau_w} = \frac{k}{\mu}\frac{(t_s - t_w)}{v_s} \tag{7.14}$$

Details of the nomenclature are shown in Fig. 7.1. v_w at the wall is zero.

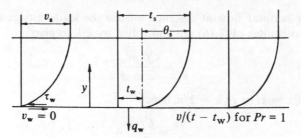

Fig. 7.1. *Velocity and temperature distributions for the simple Reynolds analogy.*

For turbulent flow, equations (7.3) and (7.12) give

$$\frac{q}{\tau} = \frac{\rho c_p(\alpha + \varepsilon_q)\,dt}{\rho(v + \varepsilon)\,dv}$$

Thus, between the free stream and wall:

$$\frac{q_w}{\tau_w} = c_p\frac{(t_s - t_w)}{v_s} \tag{7.15}$$

Equations (7.14) and (7.15) for laminar and turbulent flow are clearly identical if $Pr = 1$, i.e., if $c_p = k/\mu$, or $\mu c_p/k = 1$. Re-arranging equation (7.15) gives

$$h = \frac{q_w}{\theta_s} = \frac{\tau_w c_p}{v_s}$$

where $\theta_s = (t_s - t_w)$, and where h is the surface heat transfer coefficient.

Substituting the skin friction coefficient Cf gives

$$h = \frac{Cf}{2}\rho v_s c_p$$

or

$$\frac{h}{\rho v_s c_p} = \frac{Cf}{2} \tag{7.16}$$

This is one form of the result obtained from the simple Reynolds analogy; it gives the convection coefficient h in terms of the skin friction coefficient Cf. $h/\rho v_s c_p$ is the Stanton number St. It is the Nusselt number divided by the product of the Reynolds and Prandtl numbers. Further re-arrangement is possible; for example, considering laminar flow at distance x from the leading edge of a flat plate, both sides of (7.16) are multiplied by x/k to give

$$\frac{hx}{k} = \frac{Cf}{2} \frac{\rho v_s x c_p}{k}$$

But $c_p \mu / k = 1$, or $c_p/k = 1/\mu$, hence

$$\frac{hx}{k} = \frac{Cf}{2} \frac{\rho v_s x}{\mu}$$

or

$$Nu_x = \frac{Cf}{2} Re_x \qquad (7.17)$$

Cf may be replaced by $0.647(Re_x)^{-\frac{1}{2}}$ from equation (7.4) to give

$$Nu_x = 0.323(Re_x)^{\frac{1}{2}} \qquad (7.18)$$

for laminar flow on a flat plate. This result may be compared with equation (6.18) obtained by consideration of the integral boundary layer equations. If $Pr = 1$ in this equation then the result is

$$Nu_x = 0.332(Re_x)^{\frac{1}{2}}$$

Reynolds analogy may also be applied to flow in tubes, and for this purpose θ_s and v_s in the above analysis may be replaced by the mean values θ_m and v_m, since the velocity and temperature distributions are identical. The linear dimension is now the diameter of the tube, d. The relationship will be

$$\frac{hd}{k} = \frac{Cf}{2} \frac{\rho v_m d}{\mu}$$

or

$$Nu_d = \frac{Cf}{2} Re_d \qquad (7.19)$$

For turbulent flow in tubes, $f = 0.308(Re_d)^{-\frac{1}{4}}$ from (7.9) and $Cf = \frac{1}{4}f$ from the definition of f. Substituting for Cf in (7.19) gives

$$Nu_d = 0.038(Re_d)^{0.75} \qquad (7.20)$$

7.1.4 The Prandtl–Taylor Modification of Reynolds Analogy

The simple Reynolds analogy agreed quite well with experiment in laminar flow and also with results where $Pr = 1$ in both laminar and turbulent flow. The modification proposed by Prandtl and Taylor goes a long way to meeting the discrepancies generally found in turbulent flow when there is no restriction on Pr. A laminar sublayer is considered in addition to the turbulent boundary layer. This makes an important difference to the analysis even though the sublayer is quite thin. The fact that it is thin is also important in that it makes it possible to assume a linear temperature and velocity distribution with negligible error.

For turbulent heat and momentum exchange between the free stream and the laminar sublayer, as in Fig. 7.2, applying equation (7.15) gives:

$$\frac{q_b}{\tau_b} = \frac{c_p(t_s - t_b)}{(v_s - v_b)} \tag{7.21}$$

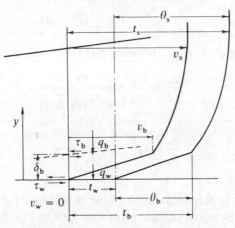

Fig. 7.2. *Velocity and temperature distributions for the Prandtl–Taylor modification of Reynolds analogy.*

In the laminar sublayer, the equations are

$$q_w = \frac{k(t_b - t_w)}{\delta_b}$$

and

$$\tau_w = \frac{\mu v_b}{\delta_b}$$

and hence

$$\frac{q_w}{\tau_w} = \frac{k(t_b - t_w)}{\mu v_b} \tag{7.22}$$

Because the velocity and temperature distributions are straight lines in the laminar region $q_w = q_b$, and $\tau_w = \tau_b$. Hence the right-hand sides of (7.21) and (7.22) are equal.

$$\therefore \qquad \frac{c_p(t_s - t_b)}{v_s - v_b} = \frac{k}{\mu} \frac{(t_b - t_w)}{v_b}$$

and

$$\frac{Pr(t_s - t_b)}{(t_b - t_w)} = \frac{v_s - v_b}{v_b}$$

If $(t_s - t_w)$ is written as θ_s, then the above may be re-arranged to give

$$\frac{(t_b - t_w)}{\theta_s} \frac{v_s}{v_b} = \frac{Pr}{1 + \dfrac{v_b}{v_s}(Pr - 1)}$$

and eliminating $(t_b - t_w)/v_b$ between this result and equation (7.22) gives

$$\frac{q_w}{\tau_w} = \frac{k\theta_s}{\mu v_s} \frac{Pr}{1 + \dfrac{v_b}{v_s}(Pr - 1)}$$

$$\therefore \qquad \frac{q_w}{\tau_w} = c_p \frac{\theta_s}{v_s} \frac{1}{1 + \dfrac{v_b}{v_s}(Pr - 1)} \tag{7.23}$$

This equation is Reynolds analogy as modified by Prandtl and Taylor. It may be noted straight away that if $Pr = 1$ in this equation, then the relationship reduces to equation (7.15), i.e., Reynolds original equation. Further, if $v_b = 0$, i.e., there is no laminar sub-layer so that flow is entirely turbulent, the equation again reduces to the original relationship. A further simplification is that if flow is all laminar, which means that $v_b = v_s$, equation (7.23) becomes

$$\frac{q_w}{\tau_w} = \frac{c_p\theta_s}{v_s Pr} = \frac{k\theta_s}{v_s \mu}$$

Equation (7.23) may now be treated in a similar manner to (7.15) by re-arranging and introducing the coefficient Cf. Thus:

$$\frac{q_w}{\theta_s} = \rho v_s c_p \frac{Cf}{2} \cdot \frac{1}{1 + \frac{v_b}{v_s}(Pr - 1)} \quad (7.24)$$

For turbulent flow on flat plates, both sides are multiplied by x/k and μ is introduced to the right-hand side to give

$$\frac{q_w x}{\theta_s k} = \frac{\rho v_s x}{\mu} \cdot \frac{c_p \mu}{k} \cdot \frac{Cf}{2} \cdot \frac{1}{1 + \frac{v_b}{v_s}(Pr - 1)}$$

$$\therefore \quad Nu_x = \frac{Cf}{2} \cdot \frac{Re_x Pr}{1 + \frac{v_b}{v_s}(Pr - 1)}$$

Also, for turbulent flow on flat plates, equations (7.5) and (7.7) are introduced to give

$$Nu_x = \frac{0.0292 Re_x^{\frac{4}{5}} Pr}{1 + 2.12 Re_x^{-\frac{1}{10}}(Pr - 1)} \quad (7.25)$$

This is the local Nusselt number. To obtain an average Nusselt number over some total length of plate, Cd from equation (7.6) may be substituted for Cf in this analysis.

An alternative to this result was suggested by Colburn,[9] in which the denominator in equation (7.25) was replaced by $Pr^{\frac{1}{3}}$. Re-arranged, this gives

$$St_x Pr^{\frac{1}{3}} = 0.0292 Re_x^{-0.2} \quad (7.26)$$

and if Cf is substituted, this gives

$$St_x Pr^{\frac{1}{3}} = \frac{Cf}{2} = J, \quad \text{(the Colburn } J\text{-factor)} \quad (7.27)$$

This result reduces to equation (7.16) when $Pr = 1$.

For turbulent flow in round tubes, equation (7.23) may be suitably modified. θ_s becomes θ_m, the temperature difference between the mean fluid temperature and the wall, and v_s similarly becomes v_m. Introducing k, μ, and the linear dimension d, gives

$$\frac{q_w d}{\theta_m k} = \frac{\rho v_m d}{\mu} \cdot \frac{c_p \mu}{k} \cdot \frac{Cf}{2} \cdot \frac{1}{1 + \frac{v_b}{v_m}(Pr - 1)}$$

$$\therefore \qquad Nu_d = \frac{Cf}{2} \cdot \frac{Re_d Pr}{1 + \frac{v_b}{v_m}(Pr - 1)}$$

Finally, equations (7.9) and (7.10) are introduced to eliminate Cf and v_b/v_m, and remembering that $f = 4Cf$, the result obtained is

$$\overline{Nu_d} = \frac{0{\cdot}0386 Re_d^{\frac{3}{4}} Pr}{1 + 2{\cdot}44 Re_d^{-\frac{1}{4}}(Pr - 1)} \qquad (7.28)$$

This is an average Nusselt number, because an average friction factor was used.

The relationships (7.25) and (7.28) agree remarkably well with experiment over a small range of Prandtl number.

EXAMPLE 7.1

Compare the heat transfer coefficients for water flowing at an average fluid temperature of 100°C, and at a velocity of 0·232 m/s in a 2·54 cm bore pipe, using the simple Reynolds analogy, equation (7.20), and the Prandtl–Taylor modification, equation (7.28). At 100°C, $Pr = 1{\cdot}74$, $k = 0{\cdot}68 \times 10^{-3}$ kW/(m K), and $v = 0{\cdot}0294 \times 10^{-5}$ m²/s.

Solution. The Reynolds number is:

$$\frac{vd}{v} = \frac{0{\cdot}232 \times 0{\cdot}0254 \times 10^5}{0{\cdot}0294} = 20{,}000$$

In the simple analogy, $\overline{Nu_d} = 0{\cdot}038 Re_d^{0{\cdot}75}$, and $Re_d^{0{\cdot}75} = 1643$

$$\therefore \qquad \overline{Nu_d} = 62{\cdot}5, \qquad \text{and} \qquad \bar{h} = \frac{62{\cdot}5 \times 0{\cdot}68 \times 10^{-3}}{0{\cdot}0254}$$

$$= 1{\cdot}675 \text{ kW/(m}^2 \text{ K)}$$

In the Prandtl–Taylor modification,

$$\overline{Nu_d} = \frac{0{\cdot}0386 Re_d^{0{\cdot}75} Pr}{1 + 2{\cdot}44 (Re_d)^{-\frac{1}{4}}(Pr - 1)}$$

$$Re_d^{\frac{1}{4}} = 3{\cdot}45$$

$$\therefore \qquad \overline{Nu_d} = \frac{0{\cdot}0386 \times 1643 \times 1{\cdot}74}{1 + (2{\cdot}44/3{\cdot}45) \times 0{\cdot}74} = 72{\cdot}4$$

$$\therefore \qquad \bar{h} = \frac{72 \cdot 4 \times 0 \cdot 68 \times 10^{-3}}{0 \cdot 0254} = 1 \cdot 937 \text{ kW}/(\text{m}^2 \text{ K})$$

The first answer is thus 13·5 per cent lower than the second, which may be assumed more correct. This solution is for flow in smooth pipes.

7.2 Dimensional Analysis of Forced Convection

Convection heat transfer is an example of the type of problem which is difficult to approach analytically, but which may be solved more readily by dimensional analysis and experiment.

The process of dimensional analysis enables an equation to be written down which relates important physical quantities, such as flow velocity and fluid properties, in dimensionless groups. The precise functional relationship between these dimensionless groups is determined by experiment.

Suppose that in a given process there are n physical variables which are relevant. These variables, which may be denoted by $Q_1, Q_2, \ldots, Q_n$, are composed of k independent dimensional quantities such as mass, length, and time. Buckingham's pi theorem[10] states that if a dimensionally homogeneous equation relating the variables may be written, then it may be replaced by a relationship of $(n - k)$ dimensionless groups.

Thus, if

$$\phi_1(Q_1, Q_2, \ldots, Q_n) = 1$$

then

$$\phi_2(\pi_1, \pi_2, \ldots, \pi_{(n-k)}) = 1$$

Each π term will be composed of the Q variables, in the general form

$$\pi = Q_1^a Q_2^b Q_3^c \ldots Q_n^x,$$

and will be dimensionless. The set of π terms will include all independent dimensionless groupings of the variables. No π term can be formed by combining other π terms. A set of equations for $a, b, c, \ldots, x$ is obtained by equating the sum of the exponents of each independent dimension to zero. This will yield k equations for n unknowns. One method of solution is to choose values for $(n - k)$ of the exponents in each term. The selected exponents must be independent, which can be shown to be true if the determinant formed from the coefficients of the others does not vanish.

An alternative procedure is to select k of the Q variables and to combine them in turn with each of the other $(n - k)$ Q variables. The selection k of the Q quantities must together involve all the independent dimensions, but they must not form a dimensionless group by themselves. Further, each of the $(n - k)$ Q variables in each π term is given the exponent 1. This facilitates the algebra, as will be seen, and is allowable since it only amounts to reducing the π term by some unknown root. Thus, if there were six Q variables and four independent dimensions, the two π terms would be:

$$\pi_1 = Q_1^{a_1}Q_2^{b_1}Q_3^{c_1}Q_4^{d_1}Q_5$$
$$\pi_2 = Q_1^{a_2}Q_2^{b_2}Q_3^{c_2}Q_4^{d_2}Q_6$$

In each π term there are therefore four simultaneous equations for the four unknown exponents.

This procedure will now be applied to forced convection. For a detailed mathematical proof of the pi theorem, the reader is referred to Langhaar.[11]

The physical variables are selected by consideration of the governing differential equations, e.g. (6.6) and (6.7) for laminar flow. The dependent variable is the convection coefficient h, and for an incompressible fluid in the absence of viscous dissipation, the independent variables are a velocity v, a linear dimension l, and the fluid properties of thermal conductivity k, viscosity μ, specific heat c_p, and density ρ. The presence of turbulence does not add any further variables. The velocity and linear dimension are normally those which define the Reynolds number for the flow, e.g., free stream velocity and distance from leading edge for flow along a flat plate, and mean velocity and diameter for flow in a tube.

The independent dimensional quantities to be used are mass M, length L, time T, temperature θ, and heat H. Heat, of course, is not independent as it has the same dimensions as kinetic energy, ML^2/T^2, but for present purposes it can be regarded as independent provided there is no transference of energy from one form to another. Heating effects due to fluid friction are consequently neglected, and the results are invalid for high speed flow. Inspection of the dimensions of the physical variables shows that when the dimensions of H and θ occur, (in h, k, and c_p), they do so in the same combination of H/θ. **Thus H/θ can be regarded as an independent dimensional quantity.**

In forced convection there are therefore seven physical variables involving four dimensional quantities. Consequently, three π terms will be obtained. Four variables which together involve all four dimensions, and which do not themselves form a dimensionless group, are v, l, k, and μ. Then h, c_p, and ρ will each appear in a separate independent π term. The π terms are

$$\pi_1 = v^{a_1} l^{b_1} k^{c_1} \mu^{d_1} h$$

$$\pi_2 = v^{a_2} l^{b_2} k^{c_2} \mu^{d_2} c_p$$

$$\pi_3 = v^{a_3} l^{b_3} k^{c_3} \mu^{d_3} \rho$$

The π_1 term may be written

$$\left(\frac{L}{T}\right)^{a_1} \left(L\right)^{b_1} \left(\frac{H}{LT\theta}\right)^{c_1} \left(\frac{M}{LT}\right)^{d_1} \frac{H}{L^2 T\theta}$$

which is dimensionless. The following equations for a_1, b_1, c_1, and d_1 are obtained:

$$L \quad : \quad a_1 + b_1 - c_1 - d_1 - 2 = 0$$

$$T \quad : \quad -a_1 - c_1 - d_1 - 1 = 0$$

$$H/\theta \quad : \quad c_1 + 1 = 0$$

$$M \quad : \quad d_1 = 0$$

It is found that $a_1 = 0$, $b_1 = 1$, $c_1 = -1$, and $d_1 = 0$. The π_1 term is thus hl/k. In a similar manner, it is found that the π_2 term is $\mu c_p/k$ and the π_3 term $\rho v l/\mu$. These groups are recognized as the Nusselt, Prandtl, and Reynolds numbers, and the result may be expressed:

$$\phi_2(Nu, Pr, Re) = 1$$

or, more usually,

$$Nu = \phi(Re, Pr) \tag{7.29}$$

since the Nusselt number contains the dependent variable h. Equation (7.29) agrees in form with Reynolds analogy, in that the Nusselt number is a function of the Reynolds and Prandtl numbers. Actual functional relationships have been determined for various fluids, geometries, and flow regimes; these may be used to predict h in similar circumstances, provided the Reynolds and Prandtl numbers fall within the same ranges. There is, of course, no restriction to the system of *units* which may be used, provided they are consistent.

Scale model testing is a valuable practical application of the use of these dimensionless relationships. By means of experiments on a model, the performance of a projected design may be estimated. The requirements are that the model must be geometrically similar to the full scale design; also that Reynolds and Prandtl numbers must be reproduced exactly. Then the flow patterns and fluid and thermal boundary layers will be correctly modelled and, consequently, the Nusselt number determined on the model will be the correct value for the real thing.

Some of the more useful results will now be summarized. It should be pointed out first, however, that the dimensional analysis just considered was based on the assumption of constant fluid properties and also that a single linear dimension was sufficient to describe the system. Both of these assumptions are invalid in certain circumstances. Viscosity is often the most temperature dependent fluid property, and a varying viscosity will have a considerable effect on the fluid boundary layer. If this is allowed for in the dimensional analysis, an additional term, such as a viscosity ratio to some power, will appear. In a result of the form of equation (7.29), fluid property values at some mean temperature are used. Consequently, when these equations are used to predict heat transfer coefficients, property values at the appropriate mean temperature must be inserted. For pipe-flow, an average or mean fluid temperature is used. If the flow across a certain section of pipe were to be thoroughly mixed, then an average fluid temperature would be obtained. It will depend on the velocity profile as well as the temperature profile. To evaluate an average heat transfer coefficient over a length of pipe, then property values at a mean of the average temperatures at the two ends must be inserted. When flow over a flat plate is being considered, a mean film temperature may be used. This is the average of the free stream fluid temperatures and the wall temperature. In addition, an average of two mean film temperatures may be used when considering an average convection coefficient over a length of plate.

When an additional linear dimension is required, as in the case of thermal boundary layer development in pipe flow, a length ratio to some power will appear in the analysis.

7.3 Empirical Relationships for Forced Convection

Some of the more important relationships are now listed.

Laminar flow in tubes. An average Nusselt number between entry and distance x from entry is given by

$$\overline{Nu_d} = 1 \cdot 86 \left[Re_d \, Pr \left(\frac{d}{x} \right) \right]^{\frac{1}{3}} \left(\frac{\mu}{\mu_w} \right)^{0 \cdot 14} \quad \text{(ref. 12)} \quad (7.30)$$

All physical properties are evaluated at the arithmetic mean bulk temperature between entry and x, with the exception of μ_w which is at the wall temperature, and the equation is valid for heating and cooling in the range of $Re \, Pr \, (d/x) > 10$. The group $Re \, Pr \, (d/x)$ is the Graetz number, Gz, and is significant in describing laminar flow heat transfer. The Nusselt number, as both an average and local value, depends heavily on the ratio (d/x) as well as on the Reynolds and Prandtl numbers. A more recent equation is:

$$\overline{Nu_d} = 1 \cdot 4 \, (Gz)^{\frac{1}{3}} \left(\frac{\mu}{\mu_w} \right)^n \quad \text{(ref. 13)} \quad (7.31)$$

where $n = 0 \cdot 05$ for heating and $1/3$ for cooling.

An equation for the local value of the Nusselt number is:

$$Nu_d = 3 \cdot 66 + \frac{0 \cdot 066 \, (Gz)}{1 + 0 \cdot 4 \, (Gz)^{\frac{2}{3}}} \quad \text{(ref. 14)} \quad (7.32)$$

As the entry length x in the Graetz number increases, the value of Nu_d approaches the constant value of $3 \cdot 66$. This compares with the value of $4 \cdot 36$ (equation 6.27), deduced from an assumed parabolic velocity distribution in fully developed flow.

In recent years, fresh analyses of laminar flow have been undertaken by means of numerical solutions of the momentum, energy and continuity equations. The work of Collins[15] has shown a close agreement with the results of Test[13], equation 7.31, and a considerable divergence from Sieder and Tate, equation 7.30.

EXAMPLE 7.2

A heat exchanger thermal wheel consists of a porous disc of available disc area A, having a porosity P. The mass flow to pass through the disc is m, and the flow area $A \times P$ consists of N holes of diameter d. The flow of gas is laminar through the holes and it is required to find the group $hA_h/(\dot{m}C_p)$ for the wheel, for a range of hole sizes, where h is the laminar heat transfer coefficient, and A_h is the area for heat transfer. (This group is the NTU, Number of Transfer Units; see also Chapter 12 and Example 12.3.)

Solution. For the wheel, $PA = N\pi d^2/4$, hence $N = 4PA/(\pi d^2)$

$$\text{and } \dot{m} = \rho PAV$$
$$\therefore \quad V = \dot{m}/(\rho PA)$$
$$\text{and } Re = \dot{m}\,d/(PA\,\mu)$$

The heat transfer area $A_h = \pi d L N$, where L is the hole length, equal to the wheel thickness. Hence

$$A_h = 4PAL/d$$

In the following program listing, the heat transfer coefficient is calculated from $Nu_d = 1\cdot86(Re\,Pr\,d/L)^{\frac{1}{3}}$, and WD = the wheel thickness L, W = the mass flow rate, CP = the fluid specific heat, DEN = the fluid density ρ, PR = the fluid Prandtl number, TK = the fluid thermal conductivity, VIS = the fluid viscosity, and D is the hole diameter, d.

Basic Program Listing and Results

```
10C      Calculation of NTU for a Thermal Wheel
20       WD=0.1
50       A=0.25
60       P=0.3
70       W=3.0
80       CP=1.02
90       DEN=0.75
100      PR=0.69
110      TK=0.037
120      VIS=0.000025
130      D=0.001
140      RE=W*D/(P*A*VIS)
150      H=1.86*TK*((RE*PR*D/WD)**0.333)/D
160      AREA=4.0*P*A*WD/D
170      TU=H*AREA/(W*CP)
```

```
180        GZ=RE*PR*D/WD
190        MOL=P*A*4.0/(3.1417*D*D)
200        LPRINT "       Matrix Hole Diameter =";
210        LPRINT USING "##.####";D;
220        LPRINT " m, Number of Holes =";
230        LPRINT USING "######";MOL
240        LPRINT "      Reynolds Number = ";
250        LPRINT USING "######.#";RE;
260        LPRINT " Graetz Number =";
270        LPRINT USING "#####.#";GZ
280        LPRINT "       Heat Transfer Coefficient =";
290        LPRINT USING "#####.#";H;LPRINT "  W/m2 K"
300        LPRINT "      Heat Transfer Area =";
310        LPRINT USING "#####.#";AREA;
320        LPRINT " m2, NTU =";
330        LPRINT USING "###.#";TU
340        LPRINT " "
350        D=D+0.0005
360        IF(D<0.0035)GO TO 140
370        STOP
```

```
Matrix Hole Diameter = 0.00100 m, Number of Holes = 95439
Reynolds Number =   1600.0 Graetz Number =    10.9
Heat Transfer Coefficient =    152.4 W/M2 K,
Heat Transfer Area =     30.0 m2, N T U = 1493.9

Matrix Hole Diameter = 0.00150 m, Number of Holes = 42439
Reynolds Number =   2400.0 Graetz Number =    24.5
Heat Transfer Coefficient =    133.1 W/M2 K,
Heat Transfer Area =     20.0 m2, N T U =  869.8

Matrix Hole Diameter = 0.00200 m, Number of Holes = 23872
Reynolds Number =   3200.0 Graetz Number =    43.5
Heat Transfer Coefficient =    120.9 W/M2 K,
Heat Transfer Area =     15.0 m2, N T U =  592.6

Matrix Hole Diameter = 0.00250 m, Number of Holes = 15278
Reynolds Number =   4000.0 Graetz Number =    68.0
Heat Transfer Coefficient =    112.2 W/M2 K,
Heat Transfer Area =     12.0 m2, N T U =  440.0

Matrix Hole Diameter = 0.00300 m, Number of Holes = 10609
Reynolds Number =   4800.0 Graetz Number =    97.9
Heat Transfer Coefficient =    105.6 W/M2 K,
Heat Transfer Area =     10.0 m2, N T U =  345.0
```

Turbulent flow in tubes. For fluids with a Prandtl number near unity, and only moderate temperature differences between the fluid and the wall, ($5°C$ for liquids, $55°C$ for gases), McAdams[16] recommends:

$$\overline{Nu_d} = 0.023(Re_d)^{0.8}(Pr)^n \qquad (7.33)$$

where $n = 0.4$ for heating, and 0.3 for cooling, and $Re_d > 10,000$. This is for fully developed flow, i.e., $(x/d) > 60$, and all fluid properties are at the arithmetic mean bulk temperature.

For both larger temperature differences and a wider range of Prandtl number:

$$\overline{Nu_d} = 0.027(Re_d)^{0.8}(Pr)^{\frac{1}{3}}(\mu/\mu_w)^{0.14} \quad \text{(ref. 12)} \qquad (7.34)$$

In this equation $0.7 < Pr < 16,700$, and all other details are as before, with μ_w taken at the wall temperature.

In many situations where tube lengths are relatively short, fully developed flow is not achieved, and the following relationship may be used:

$$Nu_d = 0.036 \, Re_d^{0.8} \, Pr^{\frac{1}{3}} \, (d/L)^{0.055} \qquad (7.35)$$

for $10 < L/d < 400$

This relationship is originally attributed to Nusselt.

The group $0.036 \, (d/L)^{0.055}$ varies between 0.0317 and 0.0259 for the range of L/d given; these figures may be compared with 0.023 and 0.027 as the numerical constants in equations (7.33) and (7.34).

Turbulent flow along flat plates. For this type of flow, Chapman recommends:

$$\overline{Nu_x} = 0.036Pr^{\frac{1}{3}}(Re_x^{0.8} - 18,700) \quad \text{(ref. 17)} \qquad (7.36)$$

This is based on a consideration of laminar flow (for which $\overline{Nu_x} = 0.664(Re_x)^{\frac{1}{2}}(Pr)^{\frac{1}{3}}$) and turbulent flow after transition at $Re_x = 400,000$, for $10 > Pr > 0.6$. Fluid properties are evaluated at the mean film temperature.

Heat transfer to liquid metals. Liquid metals are characterised by their very low Prandtl numbers. Experimental correlations are for uniform wall heat flux and constant wall temperature in turbulent flow in smooth tubes. Thus:

uniform heat flux, $\overline{Nu_d} = 0.625(Re_d Pr)^{0.4}$ (ref. 18)   (7.37)

constant wall temperature,

$$\overline{Nu_d} = 5.0 + 0.025(Re_d Pr)^{0.8} \quad \text{(ref. 19)} \qquad (7.38)$$

All properties are evaluated at the bulk temperature of the fluid, with $(x/d) > 60$, and $10^2 < (Re_d Pr) < 10^4$.

The temperature profile becomes very peaked compared with the velocity profile, when the Prandtl number is very small, as shown in Fig. 7.3.

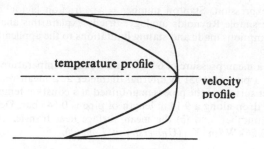

temperature profile

velocity profile

Fig. 7.3. Normalized temperature and velocity profiles for flow in a tube at very low values of Pr.

EXAMPLE 7.3

Freon at a mean bulk temperature of $-10°C$ flows at 0·2 m/s in a 20 mm bore pipe. The freon is heated by a constant wall heat flux from the pipe, and the surface temperature is 15°C above the mean fluid temperature. Calculate the length of pipe for a heat transfer rate of 1·5 kW. Use fluid properties from table A5.

Solution. At $-10°C$, $v = 0.0221 \times 10^{-5}$, $k = 72.7 \times 10^{-6}$ kw/(m K), $Pr = 4.0$, $\mu = 31.6 \times 10^{-5}$ Pa s. At $+5°C$, $\mu = 28.8 \times 10^{-5}$ Pa s.

A comparison of results using equations (7.33) and (7.34) may be obtained. $Re = 20 \times 0.2 \times 10^5/1000 \times 0.0221 = 18,100$. Therefore $Re^{0.8} = 2547$. $Pr = 4.0$, hence $Pr^{0.4} = 1.74$ and $Pr^{\frac{1}{3}} = 1.588$.

$$(\mu/\mu_w)^{0.14} = (31.6/28.8)^{0.14} = 1.013$$

From equation (7.33), $\overline{Nu_d} = 0.023 \times 2547 \times 1.74 = 102.0$

From equation (7.34), $Nu_d = 0.027 \times 2547 \times 1.588 \times 1.013 = 110.7$

Using the second result, which is 8·5 per cent larger than the first,

$$\bar{h} = 110.7 \times 72.7 \times 10^{-6} \times 10^3/20 = 0.402 \text{ kW/(m}^2 \text{ K)}$$

The pipe length required is calculated from $Q = \pi dLh(t_w - t_f)$ where t_w and t_f are the wall and fluid temperatures, hence

$$L = 1.5/(\pi \times 20 \times 10^{-3} \times 0.402 \times 15) = 3.96 \text{ m}$$

PROBLEMS

1. The expression, Stanton number $= \frac{1}{2} \times$ friction factor, may be derived from the simple Reynolds analogy. Briefly explain this analogy, discussing any assumptions made and stating limitations to the application of the above expression.

Air at a mean pressure of 6.9 bar and a mean temperature of 65.5°C flows through a pipe of 0.051 m internal diameter at a mean velocity of 6.1 m/s. The inner surface of the pipe is maintained at a constant temperature and the pressure drop along a 9.14 m length of pipe is 0.545 bar. Determine: (a) the Stanton number, and (b) the mean surface heat transfer coefficient. (Ans. 0.288, 12.56 kW/(m² K).) (*University of London*).

2. Deduce the Taylor–Prandtl equation

$$\frac{H}{F} = \frac{c\theta}{\mu}\left[\frac{1}{1 + a(Pr - 1)}\right]$$

which gives the heat transfer per unit area and time, H, in terms of the drag force per unit area, F, and in which Pr denotes the Prandtl number $c\mu/k$; the other symbols having their usual meaning. ($a = v_b/v_s$.)

Use the Taylor–Prandtl equation to show mathematically the following deductions, and explain them in simple terms:

(a) For gases the Taylor–Prandtl equation approximates closely to the Reynolds equation. (Reynolds equation is $\frac{H}{F} = \frac{c\theta}{v}$ but for liquids the divergence is considerable.)

(b) For turbulent flow the Taylor–Prandtl equation reduces to the Reynolds equation but for streamline flow it reduces to $\frac{H}{F} = \frac{k\theta}{\mu v}$.

(c) If the value of the Prandtl number is unity, then the form of the Taylor–Prandtl equation for streamline and turbulent conditions is identical.

(d) With liquids of very low thermal conductivity, the whole of the temperature drop occurs in the boundary layer. (*King's College, London*).

3. Discuss the effects of boundary layers on heat transfer by convection, and show that, if Reynolds analogy between friction and heat transfer applies,

$$\frac{h}{c_p \rho \bar{u}} = \frac{f}{2}$$

It was found during a test in which water flowed with a velocity of 2.44 m/s through a tube 2.54 cm inside diameter and 6.08 m long, that the head lost due to friction was 1.22 m of water. Estimate the surface heat transfer coefficient, based on the above analogy. For water $\rho = 998$ kg/m³, $c_p = 4.187$ kJ/(kg K). (Ans. 21.4 kW/(m² K). (*Queen Mary College, London*).

4. Air at a temperature of 115·6°C enters a smooth pipe 7·62 cm diameter, the wall of which can be maintained at a constant temperature of 15·6°C. The rate of flow of air is 0·0226 m³/sec. Estimate the length of pipe necessary if the air is to be cooled to 65·5°C, using the following assumptions: Prandtl number for air $= 0·74; f = 0·007$; velocity at boundary of sublayer is half the mean velocity in the pipe. (Ans. 12·55 m.) (*University College, London*).

5. A transformer dissipates 25 kW to cooling oil entering at 40° and leaving at 60°C. The oil is subsequently divided equally into 16 tubes in a heat exchanger. Calculate the convection coefficient of the oil in the heat exchanger tube, given: Internal tube diameter, 10 mm; oil properties: $\rho = 870$ kg/m³, $c_p = 2·05$ kJ/kg K), $\mu = 0·073$ Pa s, $Pr = 1050, k = 140 \times 10^{-6}$ kW/(m K);

$$\text{for laminar flow:} \quad Nu_d = 0·125 \, (Re_d \, Pr)^{\frac{1}{3}}$$
$$\text{for turbulent flow:} \quad Nu_d = 0·023 \, (Re_d)^{0·8} (Pr)^{\frac{1}{3}}$$

(Ans. Flow is laminar, 0·0722 kW/(m²K).) (*The City University*).

6. It is proposed to test the cooling system of an oil-immersed transformer by means of a model. The transformer dissipates 100 kW, the model is $\frac{1}{20}$ linear size, with $\frac{1}{400}$ surface area. Assuming the basic mechanism of heat transfer is forced convection in a cylindrical duct, (0·5 cm diameter on the model), determine the energy dissipation rate and the velocity in the model.

Mean temperature differences are the same in the transformer and model. Ethylene glycol is used in the model. Use $Nu_d = 0·023 \, Re_d^{0·8} \, Pr^{0·4}; \, Re = 2200$; for oil: $k = 131·5 \times 10^{-6}$ kW/(m K), $Pr = 80$; for ethylene glycol: $k = 256 \times 10^{-6}$ kW/(m K), $Pr = 80, \nu = 0·868 \times 10^{-5}$ m²/s. (Ans. 9·75 kW, 3·82 m/s.)

7. (a) Describe the following dimensionless quantities used in the study of heat transfer: Nu, Re, Pr, Gr, St, giving their physical interpretations in a form of simple ratios.

(b) Describe, using suitable formulae, what is known as Reynolds analogy.

Show that under certain conditions,

$$St = 2\tau/\rho v^2$$

(See also chapter 8.) (*University of Oxford*).

8. Air at mean conditions of 510°C, 1·013 bar, and 6·09 m/s flows through a thin 2·54 cm diameter copper tube in surroundings at 272°C.

(a) At what rate, per metre length, will the tube lose heat?

(b) What would be the reduction of heat loss if 2·54 cm of lagging with $k = 173 \times 10^{-6}$ kW/(m K) were applied to the tube? Take $N_d = 0·023(R_d)^{0·8} P^{0·33}$ with all the properties taken at the bulk air temperature. Assume the surface heat transfer coefficient from the outside of the unlagged and lagged tube to be 17·0 and 11·3 × 10^{-3} kW/(m^2 K) respectively. (Ans. 0·174 kW/m, 32 per cent.) (*University of Bristol*).

9. A 100 MW alternator is hydrogen cooled. The alternator efficiency is 98·5 per cent and hydrogen enters at 27° and leaves at 88°C. It then flows in a duct at a Reynolds number of 100,000. Calculate the mass flow rate of coolant and the duct area. For hydrogen: $c_p = 14·24$ kJ/(kg K), $\mu = 0·087 \times 10^{-4}$ Pa s. (Ans. 1·73 kg/s, 5·0 m^2.)

10. Explain and derive the simple Reynolds analogy between heat transfer and fluid friction. Outline the Prandtl–Taylor modification to the simple theory.

2·49 kg/s of air is to be heated from 15 to 75°C using a shell and tube heat exchanger. The tubes which are 3·17 cm in diameter have condensing steam on the outside and the tube wall temperature may be taken as 100°C. Specify the number of tubes in parallel and their length if the maximum allowable pressure drop is 12·7 cm of water.

Assume that $f = 0·079Re^{-\frac{1}{4}}$ and that the air has the following properties: density 1·123 kg/m^3, kinematic viscosity 1·725 × 10^{-5} m^2/s. (To solve this problem, see also chapter 13.) (Ans. 94 tubes, 3·75 m long.) (*University of Leeds*).

REFERENCES

1. Patankar, S. V. and Spalding, D. B. *Heat and Mass Transfer in Boundary Layers*, 2nd ed., International Textbook Company, Scranton, Pa. (1970).
2. Bayley, F. J., Owen, J. M. and Turner, A. B. *Heat Transfer*, Nelson (1972).
3. Reynolds, O. *Proc. Manchester Lit. Phil. Soc.*, Vol. 14, 7 (1874).
4. Reynolds, O. *Trans. Roy. Soc. Lond.*, Vol. 174A, 935 (1883).
5. Schlichting, H. *Boundary Layer Theory*, McGraw-Hill Book Company, Inc., New York (1955).
6. Knudsen, J. G. and Katz, D. L. *Fluid Dynamics and Heat Transfer*, McGraw-Hill Book Company, Inc., New York (1958).
7. Prandtl, L. *Z. Physik.*, Vol. 11, 1072 (1910).
8. Taylor, G. I. *British Adv. Comm. Aero., Reports and Mem.* ,Vol. 274, 423 (1916).
9. Colburn, A. P. *Trans. AIChE*, Vol. 29, 174 (1933).
10. Buckingham, E. *Phys. Rev.*, Vol. 4, 345 (1914).
11. Langhaar, H. L. *Dimensional Analysis and Theory of Models*, John Wiley, New York (1951).
12. Sieder, E. N. and Tate, G. E. *Ind. Eng. Chem.*, Vol. 28, 1429 (1936).
13. Test, F. L. 'Laminar flow heat transfer and fluid flow for liquids with temperature dependent viscosity', *J. Heat Trans.*, Vol. 90,

385–393 (1968).
14. Hausen, H. *VDIZ*, No. 4, 91 (1943).
15. Collins, M. W. 'Finite difference analysis for developing laminar flow in circular tubes applied to forced and combined convection', *Int. J. Num. Meth. Eng.*, Vol. 15, 381–404 (1980).
16. McAdams, W. H. *Trans. AIChE*, Vol. 36, 1 (1940).
17. Chapman, A. J. *Heat Transfer*, 3rd ed., The Macmillan Company, New York (1974).
18. Lubarsky, B. and Kaufman, S. J. NACA Tech. Note 3336 (1955).
19. Seban, R. A. and Shimazaki, T. T. *Trans. ASME*, Vol. 73, 803 (1951).

8

Natural convection

Forced convection heat transfer has now been considered in some detail. The energy exchange between a body and an essentially stagnant fluid surrounding it is another important example of convection. Fluid motion is due entirely to buoyancy forces arising from density variations in the fluid. There is often slight motion present from other causes; any effects of these random disturbances must be assumed negligible in an analysis of the process. Natural convection is generally to be found when any object is dissipating energy to its surroundings. This may be intentional, in the essential cooling of some machine or electrical device, or in the heating of a house or room by a convective heating system. It may also be unintentional, in the loss of energy from a steam pipe, or in the dissipation of warmth to the cold air outside the window of a room.

Fluid motion generated by natural convection may be laminar or turbulent. The boundary layer produced now has zero fluid velocity at both the solid surface and at the outer limit, and the profile is of the form shown in Fig. 8.1. In laminar flow natural convection

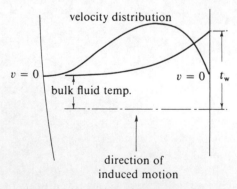

Fig. 8.1. Natural convection boundary layer on a vertical flat plate.

124

from a vertical plate, it is possible to obtain a solution of the boundary layer equations of motion and energy, if a body force term is included. This approach is limited in general application and, consequently, the method of dimensional analysis will be used.

8.1 The Body Force

Before undertaking a dimensional analysis of natural convection it is necessary to consider the nature of the body force. If ρ_s is the density of cold undisturbed fluid, ρ is the density of warmer fluid, and θ is the temperature difference between the two fluid regions, then the buoyancy force on unit volume is

$$(\rho_s - \rho)g$$

and ρ_s is related to ρ by

$$\rho_s = \rho(1 + \beta\theta)$$

where β is the coefficient of cubical expansion of the fluid. Thus the buoyancy force is

$$[\rho(1 + \beta\theta) - \rho]g = \rho g \beta \theta \qquad (8.1)$$

The independent variables on which the natural convection coefficient h depends may now be listed. A buoyancy force term would appear in the differential equation of momentum, hence β, g, and θ appear in addition to the fluid properties ρ, μ, c_p and k, and the linear dimension characteristic of the system, l. This is the dimension which would be used in the Reynolds number for a forced flow in the same direction as the natural convective flow. β and g are usually combined as a single variable βg since variation of g is unlikely.

8.2 Dimensional Analysis of Natural Convection

The procedure outlined in chapter 7 will now be followed to obtain the dimensionless groups relevant to natural convection. There are eight physical variables and five dimensional quantities, so that three π terms are expected. H and θ may not be combined to form a single dimensional quantity, since temperature difference is now an important physical variable.

Five physical variables selected to be common to all π terms are ρ, μ, k, θ, and l. These fulfil the necessary conditions. h, c_p, and βg

will each appear in a separate π term. The π terms are:

$$\pi_1 = \rho^{a_1}\mu^{b_1}k^{c_1}\theta^{d_1}l^{e_1}h$$

$$\pi_2 = \rho^{a_2}\mu^{b_2}k^{c_2}\theta^{d_2}l^{e_2}c_p$$

$$\pi_3 = \rho^{a_3}\mu^{b_3}k^{c_3}\theta^{d_3}l^{e_3}\beta g$$

After writing the necessary equations to obtain the exponents a to e in each π term, it is found that

$$\pi_1 = \frac{hl}{k}; \quad \pi_2 = \frac{\mu c_p}{k}; \quad \pi_3 = \frac{\beta g\theta\rho^2 l^3}{\mu^2}$$

The π_3 term is the Grashof number and the dimensionless relationship may be expressed as

$$\phi(Nu, Pr, Gr) = 0$$

or,

$$Nu = \phi(Gr, Pr) \qquad (8.2)$$

The Grashof number is the ratio of buoyancy force to shear force, where the buoyancy force in natural convection replaces the momentum force in forced convection. $\beta g\rho\theta$ is the buoyancy force per unit volume, therefore $\beta g\rho\theta \times l$ would be for unit area. The ratio of buoyancy to shear force per unit area is $\beta g\rho\theta l/(\mu v/l)$. But velocity is a dependent variable proportional to $(\mu/\rho l)$, hence the ratio of buoyancy to shear force becomes $\beta g\rho^2\theta l^3/\mu^2$.

Many experiments have been performed to establish the functional relationships for different geometric configurations convecting to various fluids. Generally, it is found that equation (8.2) is of the form

$$Nu = a(GrPr)^b \qquad (8.2a)$$

where a and b are constants. The product $GrPr$ is the Rayleigh number Ra. However, results are generally quoted in terms of $(GrPr)$ since it is often necessary to vary Gr at some fixed Pr. Laminar and turbulent flow regimes have been observed in natural convection, and transition generally occurs in the range $10^7 < GrPr < 10^9$ depending on the geometry.

8.3 Formulae for the Prediction of Natural Convection

Some of the more important results obtained will now be presented. These may be used for design calculations provided the system under consideration is geometrically similar and that the value of $(GrPr)$ falls within the limits specified. Generally, there are no restrictions on the use of any specific fluid. Example 8.1 shows how the formulae are used. Figure 8.2 shows the principal geometries with external flow in the direction of the arrows. For more extensive reviews of available information, see for example, refs 1 and 2. In addition, ref. 2 may be consulted for details of natural convection in enclosed spaces and natural convection effects in forced flow when the Reynolds number is very small, a situation known as combined or mixed convection.

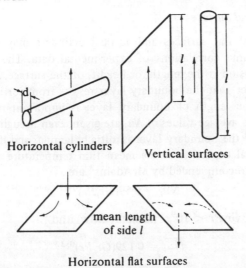

Horizontal cylinders   Vertical surfaces

mean length of side l

Horizontal flat surfaces

Fig. 8.2. Principal geometries in natural convection systems showing direction of convective flow.

8.3.1 Horizontal Cylinders

Detailed measurements indicate that the convection coefficient varies with angular position round a horizontal cylinder, but for design purposes values given by the following equations[3] are constant over the whole surface area, for cylinders of diameter d.

$$\overline{Nu_d} = 0 \cdot 525 (Gr_d Pr)^{0 \cdot 25} \tag{8.3}$$

when $10^4 < Gr_d Pr < 10^9$ (laminar flow) and

$$\overline{Nu_d} = 0 \cdot 129 (Gr_d Pr)^{0 \cdot 33} \tag{8.4}$$

when $10^9 < Gr_dPr < 10^{12}$ (turbulent flow). Below $Gr_dPr = 10^4$, it is not possible to express results by a simple relationship, and ultimately the Nusselt number decreases to a value of 0·4. At these low values of Gr_dPr the boundary layer thickness becomes appreciable in comparison with the cylinder diameter, and in the case of very fine wires heat transfer occurs in the limit by conduction through a stagnant film. Fluid properties are evaluated at the average of the surface and bulk fluid temperatures, which is the mean film temperature. If the surface temperature is unknown, a trial and error solution is necessary to find h from a known heat transfer rate.

8.3.2 Vertical Surfaces

Both vertical flat surfaces and vertical cylinders may be considered using the same correlations of experimental data. The characteristic linear dimension is the length, or height, of the surface, l. This follows from the fact that the boundary layer results from vertical motion of fluid and the length of boundary layer is important rather than its width. Again average values of Nu_l are given, even though in the case of $Gr_lPr > 10^9$ the boundary layer is initially laminar and then turbulent. With physical constants at the mean film temperature the numerical constants as recommended by McAdams[3] are

$$\overline{Nu_l} = 0·59(Gr_lPr)^{0·25}$$ (8.5)

when $10^4 < Gr_lPr < 10^9$ (laminar flow) and

$$\overline{Nu_l} = 0·129(Gr_lPr)^{0·33}$$ (8.6)

when $10^9 < Gr_lPr < 10^{12}$ (turbulent flow).

For vertical flat plates, alternative relationships are:

$$\overline{Nu_l} = \frac{0·8\,(Gr_l\,Pr)^{0·25}}{(1 + (1 + 1/Pr^{0.5})^2)^{0.25}}$$ (8.7)

(ref. 4), for Laminar flow,

and

$$\overline{Nu_l} = \frac{0·0246\,(Gr_l Pr)^{0·4}}{(Pr^{\frac{1}{6}}/(1 + 0·444\,Pr^{\frac{2}{3}}))^{\frac{2}{5}}}$$ (8.8)

(ref. 5), for turbulent flow

For moderately inclined plates, ϕ to the vertical, Gr may be multiplied by $\cos \phi$, to give the correct Grashof number based on the vertical component of the buoyancy force.

8.3.3 Horizontal Flat Surfaces

Fluid flow is most restricted in the case of horizontal surfaces, and the size of the surface has some bearing on the experimental data. The heat transfer coefficient is likely to be more variable over a smaller flat surface than a large one, when flow effects at the edges become less significant. Further, there will be a difference depending on whether the horizontal surface is above or below the fluid. Similar, though reversed, processes take place for hot surfaces facing upwards (i.e., cold fluid *above* a hot surface), and cold surfaces facing downwards (i.e., hot fluid *below* a cold surface). In either case, the fluid is relatively free to move due to buoyancy effects and be replaced by fresh fluid entering at the edges. The following relationships are generally recommended for square or rectangular horizontal surfaces up to a mean length of side (l) of 2 ft :

$$\overline{Nu_l} = 0.54(Gr_lPr)^{0.25} \qquad (8.9)$$

when $10^5 < Gr_lPr < 10^8$ (laminar flow) and

$$\overline{Nu_l} = 0.14(Gr_lPr)^{0.33} \qquad (8.10)$$

when $Gr_lPr > 10^8$ (turbulent flow). Thus turbulent flow is possible in this geometrical arrangement.

The converse arrangement is the hot surface above a cold fluid, or hot surface facing downwards, and a hot fluid above a cold surface, or cold surface facing upwards. In either case, it is obvious that convective motion is severely restricted since the surface itself prevents vertical movement. Laminar motion only has been observed, and the recommendation is

$$\overline{Nu_l} = 0.25(Gr_lPr)^{0.25} \qquad (8.11)$$

when $Gr_lPr > 10^5$. Fluid properties are again taken at the mean film temperature.

8.3.4 Additional Geometries

For two vertical parallel plates at the same temperature, convecting to air:

$$\overline{Nu_l} = 0.04 \, (Gr_l \, Pr)(d/l)^3 \qquad \text{(ref. 6)} \qquad (8.12)$$

where l is the vertical height of the plates and d is the space between them. This geometry occurs, for example, in the space between the two panels of a double hot water radiator. When the parallel plates have different temperatures, as in two panes of glass in a double-glazed window, for example:

for laminar flow: $\overline{Nu_d} = 0.18 \, (Gr_d \, Pr)^{\frac{1}{4}} \, (l/d)^{-\frac{1}{9}} (Pr)^{-\frac{1}{4}}$ \qquad (8.13)

$$\text{for } 2 \times 10^4 < Gr < 2 \times 10^5$$

for turbulent flow: $\overline{Nu_d} = 0.065(Gr_d \, Pr)^{\frac{1}{3}} \, (l/d)^{-\frac{1}{9}} (Pr)^{-\frac{1}{3}}$ \qquad (8.14)

$$\text{for } 2 \times 10^5 < Gr < 10^7 \qquad \text{(ref. 7)}$$

The vertical height is again l, the width d, and θ in Gr is the temperature difference between the plates. The relationship holds for $l/d > 3$.

For a hollow vertical cylinder with an air column

$$\overline{Nu_l} = 0.01 \, (Gr_l \, Pr) \, (d/l)^3 \qquad \text{(ref. 6)} \qquad (8.15)$$

where l is the height of the cylinder and d is the diameter. In the case of small rectangular solid objects there are generally insufficient vertical and horizontal surfaces to be treated separately, and the natural cooling of such objects in closed or open environments is a fairly common occurrence. The interaction of separate boundary layer growth means that an overall characteristic length is less than either of the actual principal dimensions, so that

$$\frac{1}{x} = \frac{1}{l_h} + \frac{1}{l_v}$$

where the characteristic length is x, and l_h and l_v are the principal horizontal and vertical dimensions, respectively. Then:

$$\overline{Nu_x} = 0.6(Gr_x \, Pr)^{0.25} \qquad \text{(ref. 8)} \qquad (8.16)$$

$$\text{for } 10^4 < GrPr < 10^9$$

8.3.5 Approximate Formulae for use with Air

A great deal of natural convection work involves air as the fluid medium and the fluid properties of air do not vary greatly over limited temperature ranges. Thus it is possible to derive simple formulae from equations (8.3) to (8.6) and (8.9) to (8.11) in which the physical properties in the Nusselt, Grashof, and Prandtl numbers are grouped together and assumed constant. From equation (8.2a)

$$h = \text{constant}\left[k^{1-b}\left(\frac{\beta g \rho^2 c_p}{\mu}\right)^b\right]\theta^b l^{3b-1}$$

$$= \text{constant} \times \theta^b l^{3b-1} \tag{8.17}$$

It will have been noted that $b = 0.25$ in laminar flow and 0.33 in turbulent flow, so that the index of l is -0.25 in laminar flow and 0 in turbulent flow. The simplified expressions become

$$h = C\left(\frac{\theta}{l}\right)^{0.25} \quad \text{in laminar flow} \tag{8.18}$$

and

$$h = C\theta^{0.33} \quad \text{in turbulent flow} \tag{8.19}$$

where the value of C, the constant, depends on the configuration and flow, and l is the characteristic dimension.

The resulting expressions for horizontal cylinders, vertical and horizontal surfaces, based on the relationships given by McAdams,[1] are:

Horizontal cylinders
d = diameter

$$h = 0.00131\left(\frac{\theta}{d}\right)^{0.25} \quad \text{laminar flow}$$

$$h = 0.00124\theta^{0.33} \quad \text{turbulent flow}$$

Vertical surfaces
l = height

$$h = 0.00141\left(\frac{\theta}{l}\right)^{0.25} \quad \text{laminar flow}$$

$$h = 0.00131\theta^{0.33} \quad \text{turbulent flow}$$

Horizontal surfaces
l = length of side

Hot, facing upwards
Cold, facing downwards

$$h = 0.00131\left(\frac{\theta}{l}\right)^{0.25} \quad \text{laminar flow}$$

$$h = 0.00152\theta^{0.33} \quad \text{turbulent flow}$$

Hot, facing downwards
Cold, facing upwards

$$h = 0.00058\left(\frac{\theta}{l}\right)^{0.25} \quad \text{laminar flow}$$

The numerical constants in these equations give h in $kW/(m^2 K)$ with θ in °C and linear dimensions in m. To determine whether flow is laminar or turbulent it is necessary to find the approximate value of $(GrPr)$ and to see to which flow regime the value corresponds, as given in earlier Sections. For this purpose take $\beta g \rho^2 c_p / \mu k = 6.4 \times 10^7$. This is multiplied by (linear dimension)$^3 \times \theta \, m^3 \, K$ to obtain the dimensionless $(GrPr)$.

EXAMPLE 8.1

An oil filled electric heating panel has the form of a thin vertical rectangle, 2 m long by 0.8 m high. It convects freely from both surfaces. The surface temperature is 85°C and the surrounding air temperature 20°C. Calculate the rate of heat transfer by natural convection, and compare the result with that obtained from the simplified formula for air.

Fluid properties at the average of surface and bulk air temperatures, 53°C, are $\beta = 1/326$, $Pr = 0.702$, $\rho = 1.087 \, kg/m^3$, $\mu = 1.965 \times 10^{-5} \, Pa \, s$ and $k = 28.1 \times 10^{-6} \, kW/(m \, K)$.

Solution. The characteristic linear dimension is the panel height, 0.8 m. The product $(Gr_l Pr)$ must first be found. $\theta = 85 - 20 = 65$.

$$Gr_l Pr = [(1/326) \times 9.81 \times 65 \times 1.087^2 \times 0.8^3/(1.965 \times 10^{-5})^2]$$
$$\times 0.702 = 2.15 \times 10^9$$

Hence the flow is turbulent, and $\overline{Nu_l} = 0.129(Gr_l Pr)^{0.33}$ may be used.

$$(Gr_l Pr)^{0.33} = 1.29 \times 10^3 \quad \therefore \quad \overline{Nu_l} = 0.166 \times 10^3$$

and

$$\bar{h} = \frac{0.166 \times 10^3 \times 28.1}{0.8 \times 10^6} = 5.85 \times 10^{-3} \, kW/(m^2 \, K)$$

Using the simplified relationship, $\bar{h} = 0.00131 \times \theta^{0.33} = 0.00131 \times 65^{0.33} = 5.26 \times 10^{-3} \, kW/(m^2 \, K)$

This represents 10.1 per cent error on the value from equation (8.6).

The convection from both sides of the panel is

$$Q = 2 \times 2 \times 0.8 \times 65 \times 5.85 \times 10^{-3} = 1.215 kW$$

using the more accurate value of $\bar{h}$.

PROBLEMS

1. Describe briefly how experimental data on heat transfer by convection obtained from small scale experiments may be applied to full-scale industrial plant, and specify the conditions which must be satisfied for this to be possible.

Define the Nusselt, Prandtl, and Grashof numbers and show that they are dimensionless. Calculate the rate of heat transfer by natural convection from the outside surface of a horizontal pipe of 15·2 cm outside diameter and 6·1 m long. The surface temperature of the pipe is 82°C and that of the surrounding air 15·6°C.

The following relations are applicable to heat transfer by natural convection to air from a horizontal cylinder; for laminar flow, when $10^4 < (GrPr) < 10^9$

$$Nu = 0.56(GrPr)^{\frac{1}{4}}$$

and for turbulent flow, when $10^9 < (GrPr) < 10^{12}$

$$Nu = 0.12(GrPr)^{\frac{1}{3}}$$

The properties of air given below, corresponding to the 'mean film temperature', i.e., 49°C, may be used.

Kinematic viscosity $v = 1.8 \times 10^{-5}$ m^2/s
Thermal conductivity $k = 0.0284 \times 10^{-3}$ kW/(mK)
Coefficient of cubical expansion $\beta = \dfrac{1}{322}$ K^{-1}

Prandtl number $Pr = 0.701$. (Ans. 1·275 kW.) (*Queen Mary College, London*).

2. The transfer of heat by natural convection from vertical planes may be calculated by using the following formula which is valid for all P, for R less than 10^9 and for N greater than 5.

$$N^4/R = 2P/(5 + 10P^{\frac{1}{2}} + 10P)$$

where $N = \dot{Q}''L/k\theta$,
$R = (\Delta\rho)gL^3\rho c_p/\mu k$,
$P = \mu c_p/k$,
and where

L = height of plane,	k = thermal conductivity
θ = temperature difference,	g = gravitational acceleration
c_p = isobaric specific heat-capacity	
μ = viscosity,	ρ = density,

$(\Delta\rho)$ = difference between density of fluid near plane and density of fluid far away,
$\dot{Q}''$ = surface density of rate of heat transfer.

Some busbars are in the form of strips which run horizontally and are ten times as high as they are thick. They are made of copper for which the resistivity is 2×10^{-8} ohm m.

They are to be designed for operation at 87·8°C in an atmosphere which is at 32·2°C and at 1·013 bar. Calculate the height of busbar for use with a current of 10,000 A. Assume that both radiation and that part of the convection which is from the top and bottom edges of the bars are negligible. (Ans. 0·35 m.) (*Queen Mary College, London*).

3. By dimensional analysis show that for natural convection of a perfect gas

$$\frac{hl}{k} = f\left\{\left(\frac{l^3 g}{v^2}\right), \left(\frac{T_s - T_0}{T_0}\right), \left(\frac{\mu c_p}{k}\right)\right\} = f(GrPr)$$

where v is the kinematic viscosity, T_s is the surface temperature and T_0 is the temperature in the bulk of the fluid. Give a brief statement of the assumptions made.

A metal plate, 0·609 m in height, forms the vertical wall of an oven and is at a temperature of 171°C. Within the oven is air at a temperature of 93·4°C and atmospheric pressure. Assuming that natural convection conditions hold near the plate, and that for this case $Nu = 0.548(GrPr)^{0.25}$, find the mean heat transfer coefficient, and the heat taken up by the air per second, per metre width. For air at 132·2°C, $k = 32.2 \times 10^{-6}$ kW/(m K), $\mu = 0.232 \times 10^{-4}$ Pa s. (Ans. 4·11 $\times 10^{-3}$ kW/(m²K), 0·195 kW/m.) (*Queen Mary College, London*).

4. A factory is heated by a bank of eight 100 mm diameter steam pipes placed under grilles in the floor. Steam at 139°C passes through the pipes and the mean air temperature in the factory is 15°C. Assuming each pipe convects freely calculate the length of the bank of pipes necessary to give 10kW of heating. (Ans. 4.19 m.)

5. A tubular heater mounted horizontally is 25 mm diameter and dissipates 0·075 kW/m length. The surrounding air temperature is 30°C. Assuming $h = 0.00127 \,(\theta/\text{diameter})^{0.25}$ kW/(m² K) with the diameter in metres, calculate the surface temperature and the value of the natural convection coefficient. (Ans. 126°C, 0·01 kW/(m² K).)

6. The surface of an electric immersion heater is equivalent to that of a given length of 10 mm diameter tube. The surface temperature must not exceed 150°C when the water temperature is 65°C and the output is 3 kW. Calculate the length of equivalent tube. (Ans. 0·44 m.)

7. The vertical wall of a building is 3 m high and it receives solar radiation at the rate of 300 W/m². Calculate the surface temperature of the wall if two-thirds of this is reradiated back to the environment, one-fifth is conducted through the wall and the remainder is dissipated by natural convection to the atmosphere at 15°C. (Ans. 43°C.)

REFERENCES

1. Hsu, S. T. *Engineering Heat Transfer*, D. Van Nostrand Company, Inc., Princeton (1963).

2. Holman, J. P. *Heat Transfer*, 3rd ed., McGraw-Hill Book Company, New York (1972).
3. McAdams, W. H. *Heat Transmission*, 3rd ed., McGraw-Hill Book Company, Inc., New York (1954).
4. Ostrach, S. *NACA TN 2635* (1952).
5. Eckert, E. and Jackson, T. *NACA TN 2207* (1950).
6. Ede, A. J. 'Advances in Free Convection', in *Advances in Heat Transfer*, Vol. 4, Academic Press, New York (1967).
7. Jacob, M. *Heat Transfer*, Vol. 1, Wiley, New York (1957).
8. King, W. J. *Mech. Eng.*, Vol. 4, 347 (1932).

9

Separated flow convection

Separation is an important characteristic of the type of flow encountered in many modern heat transfer devices. Design requirements of compactness have resulted in the rapid growth of the use of complex geometrical heat transfer surfaces, which have developed from the single tube and tube bank placed across the line of flow. A single tube or cylinder placed in a cross-flow is completely submerged in the fluid and it therefore forms an obstacle around which the fluid must flow. A boundary layer exists on the cylindrical surface with free stream velocity at its extreme and zero velocity at the wall. However, the free stream velocity increases around the front of the cylinder and at low approach velocities flow within the boundary layer also accelerates. Behind the cylinder free stream and boundary layer flow decelerates again in a more or less reverse pattern. At higher approach velocities the increased velocity around the front of the cylinder which is accompanied by a drop in static pressure is not followed by a similar increase in velocity in the boundary layer, due to the increased viscous stress at the higher velocity gradients. Thus, in the boundary layer the fluid has lost velocity before it starts to decelerate behind the cylinder and it is then opposed by a 'surplus' of static pressure which forces the boundary layer away from the surface. Separation, or break-away, results in the formation of turbulent eddies which are carried downstream behind the cylinder. Separation occurs nearer the front of the cylinder as the approach velocity increases, and occurs much more readily in flow over blunt ended obstacles.

Local heat transfer coefficients have been measured around the circumference of cylinders in cross-flow.[1] They have minimum values at the point of separation and increase forwards towards the point of stagnation, but they increase more towards the rear of the cylinder. This may be attributed to the scrubbing action of the eddies formed in that vicinity. Average values of heat transfer coefficient have also been extensively determined as these are required for design purposes. Owing to the degree of turbulence produced in a tube bank, convection coefficients are high and average values for tubes several rows back are found to be higher than for those at entry due to the action of eddies shed from the leading rows.[2]

The pattern of events in the tube bank has led to the evolution of the compact heat transfer surface which is in general a complex of finned cross-flow passages. The use of fins as a means of increasing heat transfer coefficients is discussed in chapter 12. Flow through such a system is largely composed of turbulent eddies, and even at low approach velocities a high degree of turbulence is to be found. For this reason the usual transition between laminar and turbulent flow at Reynolds numbers around 2500 does not exist, and turbulent flow has been found to persist to Reynolds numbers as low as 800.[3]

In any arrangement of this type in which high heat transfer rates may be obtained in a small space, the advantages have to be balanced against the effect of increased pressure loss on overall performance. Pressure loss is due to the total drag of the shapes involved and due to shear over the fins. It may be measured across the whole system and related to a friction coefficient by an expression similar to the equation for flow in pipes:

$$\Delta p = f_D \frac{L}{d} \rho \frac{v_m^2}{2} \tag{9.1}$$

Such a form is useful since it has been found generally that f_D can be related to the Reynolds number of flow. In the determination of f_D from Δp, the values of L, d, and v_m have to be defined in relation to the particular geometry. The symbol f_D is used to indicate that it represents essentially a drag loss rather than a loss due to viscous shear.

9.1 Relationship between Heat Transfer and Pressure Loss in a Complex Flow Sytem

In the experimental determination of the performance of complex heat transfer surfaces, Schenck[4] found that an 'experimental analogy' exists between heat transfer and friction, even though the net friction effects involved are essentially due to drag forces. Thus Fig. 9.1 shows the Colburn J-factor plotted against f_D, as defined in equation (9.1), for a wide range of different surfaces including plain fins on tubes, plain and dimpled tubes, tube and spiral fins, flattened tubes with plain grooved and wavy fins, pin fins and interrupted plate fins. This particular plot is valid for Reynolds numbers in excess of 5000.

The use of this information is illustrated in the following example.

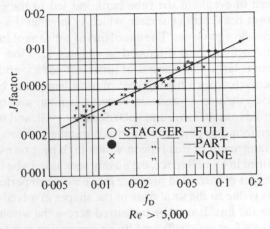

Fig. 9.1. *Relationship between the J-factor and f_D for flow in a complex system. By courtesy of H. Schenck, Jr., and The American Society of Naval Engineers, Inc.*

EXAMPLE 9.1

A compact forced convection oil cooler has a front area of $0.15 \, \text{m}^2$ and a surface area of $40 \, \text{m}^2/\text{m}^2$ frontal area. Air at 28°C enters the cooler at 30 m/s. The average temperature of the heat transfer surface is 126°C. f_D is found to be 0.1. Estimate the heat transfer performance.

Solution. From Fig. 9.1 the *J*-factor is 0.01. The *J*-factor is given by equation (7.27): $J = \overline{St}Pr^{\frac{1}{3}}$, for mean conditions where $St = \hbar/\rho v_s c_p$. At a mean temperature of 77°C, ρ for air is $0.998 \, \text{kg/m}^3$, $c_p = 1.009 \, \text{kJ/(kg K)}$, $Pr = 0.697$. Hence the heat transfer coefficient is given by

$$\frac{\hbar \times (0.697)^{\frac{1}{3}}}{0.998 \times 30 \times 1.009} = 0.01$$

$$\therefore \hbar = 0.386 \, \text{kW/(m}^2\text{ K)}$$

The heat transfer rate is $\bar{h}A\theta$, where A = (surface area/unit area) × (frontal area),

$$\therefore \bar{h}A = 0.386 \times 40 \times 0.15 \times (126 - 28)$$

$$= 227\,kW$$

9.2 Convection from a Single Cylinder in Cross Flow

Much experimental work has been done to determine the heat transfer coefficient from a single cylinder in cross flow. Investigations have included both fine heated wires and large pipes. A recent examination of available data is that of Douglas and Churchill[5] and the equation which represents their results is

$$\overline{Nu_d} = 0.46(Re_d)^{\frac{1}{2}} + 0.00128(Re_d) \tag{9.2}$$

This equation is only valid for $Re_d > 500$. Nusselt and Reynolds numbers are based on the cylinder diameter d, velocity is the free stream, or undisturbed fluid velocity, and fluid properties are evaluated at the average film temperature. Hsu[6] has proposed that for $Re_d < 500$ the following equation may be used:

$$\overline{Nu_d} = 0.43 + 0.48(Re_d)^{\frac{1}{2}} \tag{9.3}$$

Both of these equations are valid only for the simpler gases with similar Prandtl numbers, since the small Prandtl number effects are accommodated in the numerical constants. Both equations are valid in heating as well as cooling of the cylinder.

9.3 Convection in Flow across Tube Bundles

Many examples of heat transfer across tube bundles occur in industry, e.g. in cross-flow heat exchangers, and on the shell side of shell and tube heat exchangers, (see Chapter 13). It is therefore necessary to be able to predict convection coefficients in such situations.

Snyder[2] found that the local Nusselt number on tubes in cross flow achieved a constant value after the third row of tubes, and a useful correlation is that of Colburn,[7] for the average Nusselt number for all tubes, for ten or more rows of tubes in a staggered arrangement:

$$\overline{Nu_d} = 0.33\left(\frac{dG_{max}}{\mu}\right)^{0.6}(Pr)^{\frac{1}{3}} \tag{9.4}$$

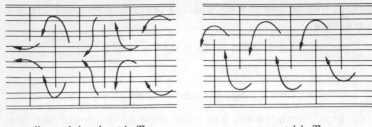

disc-and-doughnut baffles   segmental baffles

Fig. 9.2. Types of shell baffle; see also Fig. 13.2 which shows doughnut and 90° segmental baffles.

d is the tube diameter, G_{max} = mass velocity = $\rho \times v$ where v is the velocity through the smallest free-flow area between tubes, fluid properties are evaluated at mean of wall and bulk fluid temperatures, and $10 < Re_d < 40{,}000$. For a more exact dependence on Reynolds number, the numerical constant 0.33 may be replaced by 0·193 for $4000 < (dG_{max}/\mu) < 40{,}000$ and 0·0266 for $40{,}000 < (dG_{max}/\mu) < 400{,}000$. The corresponding indices of the Reynolds number are 0·618 and 0·805 (ref. 8). A much more detailed analysis for staggered and in-line tube arrangements of different spacings was carried out by Grimison.[9]

On the shell side of shell and tube heat exchangers, two relationships proposed by Donohue[10] may be used. The baffle arrangements quoted are illustrated in Fig. 9.2.

For disc-and-doughnut baffles:

$$\overline{Nu_d} = 0 \cdot 033 d_e^{0 \cdot 6} \left(\frac{dG_e}{\mu}\right)^{0 \cdot 6} (Pr)^{\frac{1}{3}} \left(\frac{\mu}{\mu_w}\right)^{0 \cdot 14} \tag{9.5}$$

For segmental baffles, the $(0 \cdot 033\, d_e^{0 \cdot 6})$ in (9·5) is replaced by 0·25. Note that $G_e = \sqrt{(G_w G_c)}$, where G_w = mass velocity through the baffle window area, and G_c = mass velocity based on flow area at the diameter of the shell. Fluid properties are evaluated at the fluid bulk temperature, with the exception of μ_w which is at the tube wall temperature. It is important to note that in using equation (9.5) all terms are dimensionless groups except for $(0 \cdot 033\, d_e^{0 \cdot 6})$. Here d_e is an equivalent diameter = $4(S_T S_V - \pi d^2/4)/(\pi d)$ where S_T = tube transverse spacing, S_V = tube vertical spacing, d = tube diameter, and d_e is in mm.

The above equations give only very basic correlations of cross-flow convection, for further information the reader is referred to Kays and London.[11]

EXAMPLE 9.2

In a shell and tube heat exchanger, the tubes are 25·4 mm diameter and are spaced at 50·8 mm centres both horizontally and vertically. Water flows at 24·6 kg/s in the shell, and the baffle window area is 0·0125 m² and the net shell area is 0·05 m². The water bulk temperature is 60°C and the tube wall temperature is 90°C. Calculate the shell side heat transfer coefficient.

Solution. Property values of water are taken from Table A5. Thus, $\mu = \rho \times v$, and at 60°C $\mu = 47·0 \times 10^{-5}$, at 90°C $\mu = 31·9 \times 10^{-5}$ Pa s. $Pr = 3·02$, $k = 651 \times 10^{-6}$ kW/(m K).

Equation 9.5 will be used. First calculate d_e, the equivalent diameter.

$$d_e = 4(S_T S_V - \pi d^2/4)/\pi d$$

$$= 4(50·8^2 - \pi \times 25·4^2/4)/\pi \times 25·4 = 104 \text{ mm}$$

$$\therefore \quad 0·033 \, d_e^{0·6} = 0·033 \times (104)^{0·6} = 0·533$$

G_w = mass velocity through baffle window = $\rho \times$ velocity. But, $\rho \times$ velocity $\times$ area = 24·6 kg/s.

$$\therefore \quad G_w = 24·6/0·0125 = 1970$$

G_c = mass velocity through the shell = 24·6/0·05 = 492

$$\therefore \quad G_e = \sqrt{(G_w G_c)} = \sqrt{(1970 \times 492)} = 984$$

$$\therefore \quad R_e = \frac{984 \times 2·54 \times 10^5}{100 \times 47·0} = 5·31 \times 10^4, \quad (Re)^{0·6} = 684$$

$$(Pr)^{\frac{1}{3}} = (3·02)^{\frac{1}{3}} = 1·445$$

$$\left(\frac{\mu}{\mu_w}\right)^{0.14} = \left(\frac{47}{31.9}\right)^{0.14} = 1.056$$

$$\therefore \quad \overline{Nu_d} = 0.533 \times 684 \times 1.445 \times 1.056 = 556.0$$

$$\therefore \quad \bar{h} = \frac{556 \times 651 \times 10^{-6}}{0.0254} = 14.3 \, kW/(m^2 \, K)$$

PROBLEMS

1. A gas is blown across two geometrically similar tube banks. In case (a) there are 10 tubes 15 mm diameter by 200 mm long, the gas velocity is 50 m/s, the gas temperature is 18°C, the tube surface temperature is 80°C and the heat transfer rate is 1.26 kW. In case (b) the ten tubes are 30 mm diameter by 400 mm long, the velocity is 30 m/s, and gas and surface temperatures are 15° and 70°C, and the heat transfer rate is 2.78 kW. With the following gas properties, determine A and B in the relationship $Nu_d = A(Re_d)^B$ for the tube banks.

(a) $k = 30 \times 10^{-6} \, kW/(mK)$, $\rho = 1.0 \, kg/m^3$ and $\mu = 2.05 \times 10^{-5} \, Pa \, s$; and (b) $k = 26 \times 10^{-6} \, kW/(mK)$, $\rho = 1.18 \, kg/m^3$ and $\mu = 1.85 \times 10^{-5} \, Pa \, s$. (Ans. $A = 0.0219$, $B = 0.81$.) (*The City University*).

2. Air at 1.5 bar and 100°C passes through a compact heat exchanger at 107 m/s. The pressure drop is 0.2 bar. Given that the values of L and d are 0.5 m and 10 mm respectively, calculate the drag loss factor f_D, the J-factor, and the heat transfer in the exchanger, assuming a flow area of 0.2 m² and a surface area of 15 m² per m² flow area. Take $c_p = 1.012 \, kJ/(kg \, K)$, $Pr = 0.692$. (Ans. $f = 0.05$, $J = 0.0072$, 630 kW, at $t_\omega = 250°C$.)

3. Hydrogen passes through a staggered bank of 200 tubes, 1.8 m long, and 25.4 mm diameter. The mass velocity is 1.5 kg/(m²s). Calculate the rate of heat transfer for a mean gas temperature of 373 K and a tube surface to gas temperature difference of 50 K. Calculate also the heat transfer rate if air at twice the mass velocity is substituted for hydrogen. (Ans. 499 kW, 70.2 kW.)

4. Carbon dioxide flows in the shell side of a shell and tube heat exchanger. There are 36 tubes 15 mm diameter by 2 m long. The shell area for flow is 0.025 m² and the baffle window area is 0.0125 m². The vertical and horizontal spacing of the tubes is 22.5 mm between centres. The mass flow of carbon dioxide is 0.6 kg/s at a mean temperature of 400 K. The mean tube surface temperature is 300 K. Calculate the convective heat transfer coefficient on the shell side of the tubes and the heat transfer rate. (Ans. 0.168 kW/(m² K), 56.8 kW.)

REFERENCES

1. Schmidt, E. and Wenner, K. *Forschung, Gebiete Ingenieurw.*, Vol. 12, 65 (1933).
2. Snyder, N. W. *Chem. Eng. Progr.*, Symposium Series, Vol. 49, No. 5, 11 (1953).
3. Schenck, H. Jnr. *Heat Transfer Engineering*, Longmans, Green and Co. Ltd. (1960).
4. Schenck, H. Jnr. *J. Amer. Soc. Naval Eng.*, Vol. 69, 767 (1957).
5. Douglas, M. J. M. and Churchill, S. W. *Chem. Eng. Propr.*, Symposium Series, Vol. 52, No. 18, 23 (1956).
6. Hsu, S. T. *Engineering Heat Transfer*, D. Van Nostrand Company, Inc., Princeton (1963).
7. Colburn, A. P. *Trans. AIChE*, Vol. 29, 174 (1933).
8. Holman, J. P. *Heat Transfer*, McGraw-Hill Book Company, Inc., New York (1981).
9. Grimison, E. D. *Trans. ASME*, Vol. 59, 583 (1937).
10. Donohue, D. A. *Ind. Eng. Chem.*, Vol. 41, 2499 (1949).
11. Kays, W. M. and London, A. L. *Compact Heat Exchangers*, McGraw-Hill Book Company, Inc., New York (1964).

10

Convection with phase change

Convection processes with phase change are of great importance, particularly those involving boiling and condensing in the fluid phase. Such processes occur in steam power plant and in chemical engineering plant. Convection in the liquid to solid phase change is also of importance, as for example in metallurgical processes, but this cannot be considered here.

10.1 Description of Condensing Flow

Two types of condensation are recognized, in which the condensing vapour forms either a continuous film of liquid on the solid surface, or a large number of droplets. Film condensation is the more common; drop formation occurs generally in an initial transient stage of condensing flow, or if for any reason the surface is unwettable. A condensing vapour generally forms droplets around nuclei of minute solid particles, and these droplets merge into a continuous film as they grow in number and size. The film then flows under the action of gravity so that the process may continue. As condensation depends on conduction of heat away through the solid surface, the growth of a liquid film will impede the condensation rate. Condensation is also impeded if a non-condensable gas is mixed with the vapour, since the concentration of gas tends to be greater at the surface as the vapour changes its phase, and this acts as a thermally insulating layer. It is thus desirable to prevent the film growing in thickness, and for this reason horizontal tubes are most commonly used as the condensing surface. Cold water flows inside the tube whilst the vapour condenses outside. The tubes are staggered vertically to prevent too great a build-up of film on the lower tubes as liquid drips off the upper ones. In comparison with the horizontal tube a vertical tube or flat surface

144

will allow the liquid film to grow in thickness considerably, and the average heat transfer rate per unit area is somewhat smaller than for the horizontal tube.

10.2 A Theoretical Model of Condensing Flow

Nusselt proposed an analysis of condensation in 1916.[1] This was applied first to a vertical surface and the same mechanism was then extended to the horizontal tube. The results agree well with experiment. The analysis of the vertical surface will be given here to illustrate the method, and the reader may refer to the literature for the more lengthy analysis of the horizontal tube.[2,3]

Certain simplifying assumptions are made in the analysis. The film of liquid formed flows down the vertical surface under the action of gravity and flow is assumed everywhere laminar. Only viscous shear and gravitational forces are assumed to act on the fluid, thus inertial and normal viscous forces are neglected. Further, there is no viscous shear between the liquid and vapour phases, so there is no velocity gradient at the phase interface. (The temperature of the surface is assumed constant at t_w and the vapour is saturated at temperature $t_{sat.}$). The mass flow rate down the surface increases with distance from the top; this increase is associated with the amount of fluid condensing at any chosen point. The model to be considered is shown in Fig. 10.1. The velocity profile is of the form shown, with $v_x = 0$ at the surface, and $(\partial v_x/\partial y)_{y=\delta} = 0$ at the liquid–vapour interface.

Assuming that the vertical surface has unit width, it is necessary to consider an element of fluid $dx\,dy$ and unit depth, at a distance x from the top of the plate. The body force on this element is $\rho g\,dx\,dy$. The shear stress at y is

$$\tau_y = \mu\frac{\partial v_x}{\partial y}$$

The shear stress at $y + dy$ is

$$\tau_{y+dy} = \tau_y + \frac{\partial \tau_y}{\partial y}dy = \mu\frac{\partial v_x}{\partial y} + \mu\frac{\partial^2 v_x}{\partial y^2}dy$$

These shear stresses act over an area $1 \times dx$. Balancing the forces

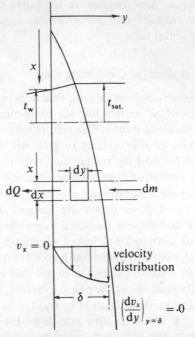

Fig. 10.1. *Condensation on a vertical surface.*

gives

$$\rho g \, dx \, dy = (\tau_y - \tau_{y+dy}) \, dx = -\mu \frac{\partial^2 v_x}{\partial y^2} \, dx \, dy$$

$$\therefore \quad \frac{d^2 v_x}{dy^2} = -\frac{\rho g}{\mu}$$

and on integration,

$$v_x = -\frac{\rho g y^2}{2\mu} + C_1 y + C_2$$

The boundary conditions are that $v_x = 0$ at $y = 0$ and $dv_x/dy = 0$ at $y = \delta$, the thickness of the film. Hence

$$C_2 = 0, \quad \text{and} \quad -\frac{\rho g \delta}{\mu} + C_1 = 0$$

The equation for v_x is thus

$$v_x = -\frac{\rho g}{\mu}\left(\frac{y^2}{2} - y\delta\right) \tag{10.1}$$

The mass flow at x can then be obtained by integrating over the film thickness δ. Thus

$$m = \int_0^\delta \rho v_x \, dy = \int_0^\delta -\frac{\rho^2 g}{\mu}\left(\frac{y^2}{2} - y\delta\right) dy$$

$$= -\frac{\rho^2 g \delta^3}{6\mu} + \frac{\rho^2 g \delta^3}{2\mu} = \frac{\rho^2 g \delta^3}{3\mu}$$

But δ is a function of x, and

$$\frac{dm}{dx} = \frac{\rho^2 g \delta^2}{\mu} \cdot \frac{d\delta}{dx} \qquad (10.2)$$

Next, the heat transfer, dQ, resulting from the condensation of an element of matter, dm, may be considered. This quantity of energy is conducted across the film to the wall, so by Fourier's law,

$$dQ = \frac{k \, dx(t_{\text{sat.}} - t_w)}{\delta} = \frac{k \, dx\theta_w}{\delta} \qquad (10.3)$$

where dx is the area of the element of surface of unit depth. dQ may also be expressed as dmh_{fg}, assuming the vapour is saturated and there is no undercooling of liquid. From these relationships, dm may be expressed as

$$dm = \frac{k\theta_w dx}{h_{fg}\delta}$$

or

$$\frac{dm}{dx} = \frac{k\theta_w}{h_{fg}\delta} \qquad (10.4)$$

Equations (10.2) and (10.4) may be combined to give

$$\frac{\rho^2 g \delta^2}{\mu} \frac{d\delta}{dx} = \frac{k\theta_w}{h_{fg}\delta}$$

This result may be integrated between the top of the surface down to x to give

$$\frac{\rho^2 g \delta^4}{4\mu} = \frac{k\theta_w x}{h_{fg}}$$

or

$$\delta = \left(\frac{4\mu k\theta_w x}{h_{fg}\rho^2 g}\right)^{\frac{1}{4}} \qquad (10.5)$$

This is the relationship between film thickness and distance x from the top of the surface. From equation (10.3) a convection coefficient may be obtained as

$$h_x = \frac{dQ}{dx\theta_w} = \frac{k}{\delta}$$

and hence

$$Nu_x = \frac{h_x x}{k} = \frac{x}{\delta} = \left(\frac{h_{fg}\rho^2 g x^3}{4\mu k\theta_w}\right)^{\frac{1}{4}}$$

Thus the local Nusselt number may be written as

$$Nu_x = 0.706\left(\frac{h_{fg}\rho^2 g x^3}{\mu k\theta_w}\right)^{\frac{1}{4}} \tag{10.6}$$

An average Nusselt number is then obtained by integrating h_x from 0 to x and dividing the result by the area $x \times$ unit depth, to give

$$\overline{Nu}_x = \frac{4}{3}Nu_x = 0.943\left(\frac{h_{fg}\rho^2 g x^3}{\mu k\theta_w}\right)^{\frac{1}{4}} \tag{10.7}$$

The analysis on the horizontal tube of diameter d yields a similar expression for the average Nusselt number, thus

$$\overline{Nu}_d = 0.725\left(\frac{h_{fg}\rho^2 g d^3}{\mu k\theta_w}\right)^{\frac{1}{4}} \tag{10.8}$$

EXAMPLE 10.1

Steam at 0·25 bar absolute condenses on 30 mm diameter horizontal tubes which have a surface temperature of 40°C. Calculate the average heat transfer coefficient.

Solution. The saturation temperature is 65°C at which $h_{fg} = 2345.7$ kJ/kg. The mean film temperature (at which liquid fluid properties are taken) is 53°C. Hence $\rho = 986$ kg/m³, $\mu = 526 \times 10^{-6}$ Pa s, and $k = 646 \times 10^{-6}$ kW/(m K). $\theta_w = (t_{sat} - t_w) = 25$°C. Equation (10.8) gives

$$\overline{Nu}_d = 0.725\left(\frac{2345.7 \times 986^2 \times 9.81 \times 0.03^3}{526 \times 646 \times 10^{-12} \times 25}\right)^{\frac{1}{4}}$$

$$= 0.725 \times (712 \times 10^8)^{\frac{1}{4}}$$

$$= 0.725 \times 517 = 375.0$$

$$\bar{h} = 375.0 \times \frac{k}{d} = \frac{375.0 \times 646 \times 10^{-6}}{30 \times 10^{-3}}$$

$$= 8.08 \text{ kW/(m}^2 \text{ K)}$$

Equation (10.7) for a vertical surface may be applied to a vertical tube provided the diameter is not small, when the liquid film becomes two-dimensional, and hence it is possible to compare the relative merits of horizontal and vertical tubes. Thus

$$\frac{\overline{Nu_d}}{\overline{Nu_x}} = \frac{0.725}{0.943}\left(\frac{d^3}{x^3}\right)^{\frac{1}{4}} = \frac{h_d d}{h_x x}$$

$$\therefore \quad \frac{h_d}{h_x} = 0.770\left(\frac{x}{d}\right)^{\frac{1}{4}}$$

If (x/d) is 75, say, it follows that $h_d = 2.26\,h_x$. Thus over twice the fluid is condensed with the tubes arranged horizontally, h_x being the coefficient for the vertical tube. Condensation inside a tube is a process of some interest since it occurs in refrigeration and heat pump condensers, but it is a process of considerable complexity and is conveniently described by empirical relationships. For low vapour velocities Chato[4] gives:

$$Nu_d = 0.555\left[\frac{d^3\rho_1(\rho_1 - \rho_v)\,g\,(h_{fg} + 0.68\,C_{pl}(t_{sat} - t_w))}{k_1\mu_1(t_{sat} - t_w)}\right]^{\frac{1}{4}} \quad (10.9)$$

for short tubes at Re (vapour) $< 35{,}000$.

Suffices l and v refer to liquid and vapour states respectively. The subject of condensing flow inside tubes is described further by Akers, Deans and Crosser.[5]

For more advanced topics on condensation the reader is referred to the literature. It is not possible to consider in this introductory text the effects of turbulence in the liquid film,[3] velocity of the condensing vapour[6] or, superheat.[3]

10.3 Boiling Heat Transfer

Heat transfer to boiling liquids is a subject at present under intensive study. It is of paramount importance in the power generation industry. Several fairly well defined regimes of heat transfer are now recognized, and values of heat transfer coefficient associated with each have been measured.

Thus when there is a free liquid surface above the heated surface, the regime is known as *pool boiling*, and *sub-cooled* boiling occurs when the bulk liquid temperature is below the saturation value. As the temperature rises to saturation, *saturated boiling* occurs, increasing in intensity as the surface temperature rises to give *bulk boiling*. The term *nucleate* boiling is associated with these regimes as bubbles leave nucleation sites, leading to *film boiling* as bubbles completely cover the surface.

A simple experiment involving an electrically heated wire immersed in water illustrates the simpler boiling mechanisms.[6] The variation of heat flux with the difference in temperature between the wire and liquid has been observed by numerous investigators and the general form of the result is shown in Fig. 10.2. As the wire warms up initially heat transfer is by natural convection. As the wire temperature reaches a few degrees in excess of the saturation temperature streams of tiny bubbles will be observed to leave the surface of the wire. These bubbles are produced at nucleation sites, since a minor roughness of the surface is necessary for the bubble to form. Higher temperatures are found to be necessary for nucleation to begin if the surface is made especially smooth. Part 1–2 of the curve in Fig. 10.2 is natural convection, and this becomes steeper in region 2–3 as boiling proceeds. This initial boiling is known as nucleate boiling. The heat transfer rate is significantly

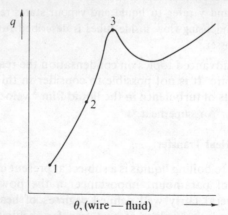

Fig. 10.2.   The boiling curve, after Farber and Scorah.[7]

improved by the stirring action of the bubbles. Bubble formation becomes increasingly energetic as point 3 is approached. At this point the bubbles tend to merge together to form a continuous vapour enclosure round the wire. When this happens nucleate boiling gives way to film boiling and there is a reduction in heat flux due to the thermally insulating effect of the vapour. This situation leads to a rapid increase in wire temperature and possible melting, unless the current input is quickly reduced. Once film boiling is safely established, the heat flux will again increase with temperature until the wire melts, the mechanism here being convection and radiation through the vapour.

Many useful calculations on boiling may be made from the Rohsenow correlation[8] which is in terms of the difference in temperature between the surface and the fluid saturation value and the heat flux per unit area, for a number of surface/liquid combinations

$$\frac{c_{pl}\theta}{h_{fg}Pr_l^{1.7}} = C_{sf}\left[\frac{Q/A}{\mu_l h_{fg}}\sqrt{\left(\frac{\sigma}{g(\rho_1 - \rho_v)}\right)}\right]^{0.33} \quad (10.10)$$

where

c_{pl} = specific heat of saturated liquid

h_{fg} = enthalpy of vapourisation

Pr_l = Prandtl number of saturated liquid

μ_l = viscosity of saturated liquid

ρ_1 = density of saturated liquid

ρ_v = density of saturated vapour

σ = surface tension of liquid vapour interface

θ = heated surface saturation temperature difference

Q/A = heat flux per unit area

g = gravitational acceleration

C_{sf} = experimental constant

The value of C_{sf} is 0·013 for water–copper and water–platinum, and 0·006 for water–brass. The equation is dimensionless, so any system of units may be used without correction.

The use of this correlation may be extended to flow in tubes, when Rohsenow and Griffith[9] recommend that a convective heat flux may be calculated from (7.34) and added to that from (10.10) to obtain a total heat flux for the boiling flow.

Boiling processes may be further sub-divided when considering the flow of fluid vertically in a tube. The process may be associated with the type of flow.[10] Various flow regimes are shown in Fig. 10.3 These are: sub-cooled liquid flow, 'frothy' or 'bubbly' flow at low dryness fraction, 'churn' or 'slug' flow in which slugs of vapour appear, annular or climbing film flow, fog or dispersed liquid flow, and finally dry wall flow at the saturated steam condition. Associated boiling processes are tabulated in Fig. 10.3. Sub-cooled nucleate and film boiling are examples of local boiling. There is no overall production of vapour; this is condensed in the main bulk of the fluid after being produced at the wall of the tube. Very high convection coefficients result because of the activity at the wall, and this heat transfer mechanism is finding application in other situ-

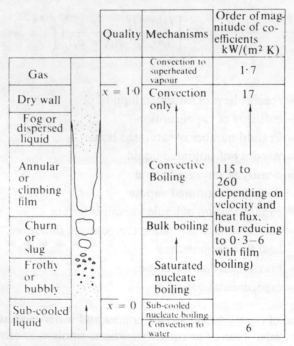

	Quality	Mechanisms	Order of magnitude of coefficients $kW/(m^2 K)$
Gas		Convection to superheated vapour	1·7
Dry wall	$x = 1\cdot0$	Convection only	17
Fog or dispersed liquid			
Annular or climbing film		Convective Boiling	115 to 260 depending on velocity and heat flux. (but reducing to 0·3−6 with film boiling)
Churn or slug		Bulk boiling	
Frothy or bubbly		Saturated nucleate boiling	
Sub-cooled liquid	$x = 0$	Sub-cooled nucleate boiling	
		Convection to water	6

Fig. 10.3.   Flow and boiling regimes in a vertical heated tube. From data of Firman, Gardener, and Clapp.[10] *By courtesy of the Institution of Mechanical Engineers*

ations where a high convection coefficient is valuable. Saturated nucleate boiling occurs when the bulk fluid temperature has reached the saturation value, and is therefore associated with flow at low dryness fraction. This mechanism persists into the slug flow regime when it is termed bulk boiling. When, with the increasing velocities, annular flow is established, convective heat transfer between the annulus of liquid and the core of vapour takes place and the nucleate process tends to be suppressed. This is known as convective boiling. Initially, the vapour core is thought to be fairly dry, but with accelerated flow the liquid annulus is entrained as a dispersed spray or fog in the core. Once the liquid phase has left the tube wall, as in the dry wall region, the heat transfer coefficient drops rapidly. The mechanism is by convection and by conduction to individual droplets impinging on the wall. Finally, when the steam becomes superheated, heat transfer is by convection only. Film boiling is avoided in the foregoing as far as possible. It occurs with excessive heat fluxes and results in drastic reductions in the boiling coefficient and very high metal temperatures. The order of magnitude of the heat transfer coefficients associated with the type of flow and mechanism of heat transfer are also shown in Fig. 10.3. It will be observed that the coefficients vary over a considerable range.

It will be appreciated from what has been said so far that boiling heat transfer is a complex subject and to take the subject any further is beyond the scope of this text. Working formulae and procedures exist in the literature for the determination of boiling coefficients for design purposes, and the reader may refer to Bagley[11] for a recent statement from the boiler industry, and to Jakob[12] and to Hsu[3] for more comprehensive treatments of the subject.

EXAMPLE 10.2
Using the Rohsenow equation, calculate the heat transfer coefficient for boiling when water boils at atmospheric pressure in a copper pan with the copper surface at 120°C, and compare with the convection coefficient for water flowing in a 40 mm diameter tube at 1 m/s under the same conditions, using equation (7.34). Use $C_{sf} = 0.013$, $c_{p1} = 4.216 \, kJ/(kg \, K)$, $h_{fg} = 2256.7 \, kJ/kg$, $Pr_1 = 1.74$, $\mu_1 = 279 \times 10^{-6} \, Pa \, s$, $\rho_1 = 957 \, kg/m^3$, $\rho_v = 0.598 \, kg/m^3$, $\sigma = 0.0587 \, N/m$. At a mean temperature of 110°C, $\rho_1 = 950 \, kg/m^3$, $\mu_1 = 252 \times 10^6 \, Pa \, s$, $Pr = 1.56$; $k = 684 \times 10^{-6} \, kW/(m \, K)$ and at 120°C $\mu_1 = 230 \times 10^{-6} \, Pa \, s$.

Solution. The Rohsenow equation will give Q/A from which h may be found. Thus:

$$\frac{4 \cdot 216 \times 20}{2256 \cdot 7 \times (1 \cdot 74)^{1 \cdot 7}} = 0 \cdot 013 \left[\frac{Q/A \times 10^6}{279 \times 2256 \cdot 7} \times \sqrt{\left(\frac{0 \cdot 0587}{9 \cdot 81(957 - 0 \cdot 598)} \right)} \right]^{0 \cdot 33}$$

$$\therefore \qquad Q/A = 358 \cdot 0 \text{ kW/m}^2$$

and $h = (Q/A)/\theta = 358 \cdot 0/20 = 17 \cdot 9$ kW/(m² K). From equation (7.34),

$$\frac{\bar{h}d}{k} = 0 \cdot 027 \times \left(\frac{950 \times 1 \times 40 \times 10^6}{252 \times 10^3} \right)^{0 \cdot 8} \times (1 \cdot 56)^{\frac{1}{3}} \times \left(\frac{252}{230} \right)^{0 \cdot 14}$$

$$= 442$$

$$\therefore \qquad \bar{h} = \frac{442 \times 684 \times 10^3}{10^6 \times 40} = 7 \cdot 57 \text{ kW/(m}^2 \text{ K)}$$

PROBLEMS

To solve Question 1 see also chapter 12.

1. An air heater consists of horizontal tubes 30 mm diameter and 23 mm bore arranged in vertical banks of twenty. Air passes inside the tubes and is heated from 32°C to 143°C by saturated steam at 180°C which passes over the tubes. The mean air velocity is 23 m/s and the air flow 3·82 kg/s. Calculate the number and length of tubes required. The heat transfer coefficient for saturated steam to tube surface (h_{sv}) can be found from

$$h_{sv} = 0 \cdot 725 \left(\frac{k_c^3 \rho_c^2 g h_{fg}}{N d \mu_c \Delta t} \right)^{\frac{1}{4}} \text{ kW/(m}^2 \text{ K)}$$

where the suffix c denotes condensate properties evaluated at the saturation temperature, g is the gravitational acceleration in m/s², N is the number of horizontal tubes in a vertical bank, d is the outside diameter in m, t is the temperature difference between the saturated vapour and the tube surface and may be assumed to be 11°C. The other symbols have their usual meaning. (Ans. 400 tubes, 2·34 m.) (*University of Glasgow*).

2. Water flows in a 0·8 cm bore copper tube at a Reynolds number of 10,000. The saturation temperature is 105°C and the wall temperature 130°C. Calculate the boiling heat flux using the Rohsenow equation and hence the total heat flux. Use the following property values: $\sigma = 0 \cdot 0525$ N/m, $h_{fg} = 2244$ kJ/kg, $\rho_l = 954$ kg/m³, $\rho_v = 0 \cdot 71$ kg/m³, $c_{pl} = 4 \cdot 23$ kJ/(kg K), $Pr_l = 1 \cdot 64$, $\mu_l = 265 \times 10^{-6}$ Pa s, $\mu_w = 230 \times 10^{-6}$, $k = 687 \times 10^{-6}$ kW/(m K). (Ans. 956 kW/m², 979 kW/m².)

3. Describe the 'Farber–Scorah Boiling Curve' together with the mechanism of heat transfer relating to each section of the curve. Discuss the following

topics in relation to the heat transfer to a fluid in which nucleate boiling occurs:
  (a) Temperature distribution in the fluid;
  (b) The nature of the heating surface:
  (c) The operating pressure.

  (*University of Leeds*).

4. Steam is being condensed on flat vertical surfaces. If the drag on the steam side of the condensate film can be neglected, derive an expression for the local and mean heat transfer coefficient on the surface.
  Discuss the assumptions which you make in the derivation.
  If the surfaces are parallel and steam enters the space between two surfaces at the top, show how you would correct the derivation for the drag of the flowing steam on the condensate film. (*University of Leeds*).

5. Outline the Nusselt theory of film condensation, indicating the steps which lead to the following formula for the average surface heat transfer coefficient h_m during the condensation of a saturated vapour on a plane vertical surface:

$$Nu_m = \frac{h_m L}{K} = 0.943 \left(\frac{\rho^2 g L^3 h_{fg}}{\mu K \Delta T}\right)^{\frac{1}{4}}$$

L is the height of the surface, g the acceleration due to gravity, h_{fg} the enthalpy of evaporation, ΔT the difference between the temperatures of the vapour and the surface and ρ, μ, and K are respectively the density, absolute viscosity, and thermal conductivity of the condensate at the saturation temperature.
  Saturated steam at 149°C is to be condensed in a cylinder of diameter 1·217 m and length 0·305 m, having its axis vertical. The curved wall is maintained at 10°C by external coolant and no condensation takes place on the two horizontal surfaces. The steam is fed in through a pipe in the top surface of the cylinder.
  Determine the initial average surface heat transfer coefficient, and estimate the time taken to fill the container with water which may be assumed to remain at 149°C. (Ans. 4·85 kW/(m² K), 0·976 h.) (*University of Cambridge*).

REFERENCES

1. Nusselt, W. Z. *d. Ver. deutsch. Ing.*, Vol. 60, 541 (1916).
2. Nusselt, W. Z. *d. Ver. deutsch. Ing.*, Vol. 60, 569 (1916).
3. Hsu, S. T. *Engineering Heat Transfer*, D. Van Nostrand Company, Inc., Princeton (1963).
4. Chato, J. C. *J. Am. Soc. Refrig. Air Cond. Eng.*, Feb., 52 (1962).
5. Akers, W. W., Deans, H. A. and Crosser, O. K. *Chem. Eng. Progr.*, Symposium Series, Vol. 55, No. 29, 171 (1959).
6. Carpenter, F. G. and Colburn, A.P. 'General Discussion on Heat Transfer', *I. Mech. E. London* (1951).
7. Farber, E. A. and Scorah, R. L. *Trans. ASME.*, Vol. 70, 369 (1948).
8. Rohsenow, W. M., *Trans. ASME*, Vol. 74, 969 (1952).
9. Rohsenow, W. M. and Griffith, P. *AIChE–ASME Heat Transfer Symposium*, Louisville, Ky (1955).

10. Firman, E. C., Gardner, G. C., and Clapp, R. M. *I. Mech. E. Symposium on Boiling Heat Transfer*, Manchester, Review Paper 1 (1965).
11. Bagley, R. *I. Mech. E, Symposium on Boiling Heat Transfer*, Manchester, Paper 13 (1965).
12. Jakob, M. *Heat Transfer*, Vol. 2, John Wiley, New York (1957).

11

Extended surfaces

Convection from a solid surface to a surrounding fluid is limited by the area of that surface. It would seem reasonable, therefore, that if the surface area could be extended, then a gain in total heat transfer would be achieved. This is done by adding fins to the surface. Heat transfer is then by conduction along the fin, and by convection from the surface of the fin. It is likely that the convection coefficient of the basic surface will be altered by the addition of fins, due to the new flow pattern involved and the fact that the temperature of the fin surface will not be uniform. Though the *average* surface temperature is reduced by the addition of fins, the total heat transfer is increased. In the treatment that follows it is assumed that the convection coefficient is known. The Nusselt numbers of finned surfaces may be determined experimentally.

There are various types of fin, the most common being the straight fin, the spine, and annular fin. The straight fin is rectangular in shape and generally of uniform cross-section, and the spine is simply a short thin rod protruding from the surface. Annular fins are often found if the primary or basic surface is cylindrical. Examples are to be found in heat exchangers and air-cooled petrol engines. Extended surface nuclear fuel cans are shown in Fig. 11.1. These are both straight and spiral in form.

Only the straight fin and spine will be considered here in detail. Fins of non-uniform cross-section and annular fins are more complex mathematically, and the reader is referred elsewhere for details.[1,2,3]

11.1 The Straight Fin and Spine

These are shown in Fig. 11.2. The straight fin has length L, and height l (from root to tip). These definitions are used whatever the actual orientation of the fin may be. In developing the theory of h

157

transfer in a fin it is assumed that the thickness, or diameter of the spine, is small compared with the length. Conduction along the fin may then be assumed to be one-dimensional. The conduction and convection heat transfers involved are shown in Fig. 11.3. Two important dimensions of fins are their area of cross-section A, and their perimeter P. In the straight fin it is convenient to assume that a is small compared with L. Thus:

$$\text{Straight fins} \quad A = La, \qquad P = 2L$$
$$\text{Spines} \qquad\quad A = \tfrac{1}{4}\pi d^2, \qquad P = \pi d$$

Consider an element of a fin or spine as shown in the figure. Conduction into the element at x is Q_x. This must be equal to the sum of the conduction out of the element at $x + dx$ and the

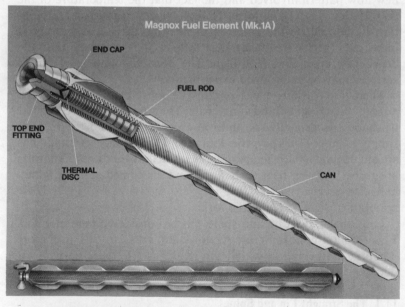

Fig. 11.1 *Magnox fuel element, as used in the U.K. gas cooled reactor programme. Note the external extended surface configuration. This illustration is reproduced by courtesy of Nuclear Fuels, plc.*

convection from the surface of the edge of the element. Thus

$$Q_x = -kA\frac{dt}{dx}$$

$$Q_{(x+dx)} = -kA\frac{dt}{dx} - kA\frac{d^2t}{dx^2}dx$$

$$Q_h = hP\,dx(t - t_s)$$

and

$$Q_x = Q_{(x+dx)} + Q_h$$

$$\therefore \quad -kA\frac{d^2t}{dx^2}dx + hP\,dx(t - t_s) = 0$$

$$\therefore \quad \frac{d^2t}{dx^2} - \frac{hP}{kA}(t - t_s) = 0$$

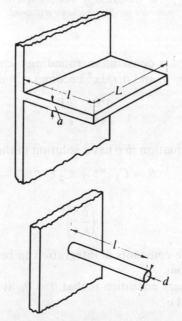

Fig. 11.2. The straight fin and the spine.

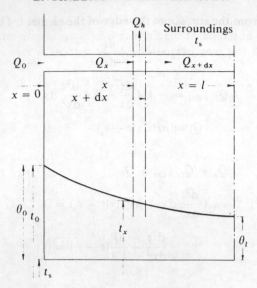

Fig. 11.3. Heat transfer from an extended surface. This diagram implies heat transfer from the fin to the surroundings. For the reverse case an inverted temperature profile would be obtained.

Since t_s is assumed a constant surroundings temperature, $(t - t_s)$ may be replaced by θ, and d^2t/dx^2 becomes $d^2\theta/dx^2$.

$$\therefore \quad \frac{d^2\theta}{dx^2} - \frac{hP}{kA}\,\theta = 0$$

This differential equation in θ has a solution of the form:

$$\theta = C_1\,e^{mx} + C_2\,e^{-mx} \tag{11.1}$$

where

$$m = \left(\frac{hP}{kA}\right)^{\frac{1}{2}} \tag{11.2}$$

and C_1 and C_2 are constants of integration to be determined from boundary conditions.

The first boundary condition is that $\theta = \theta_0$ at $x = 0$. Therefore, from equation (11.1):

$$\theta_0 = C_1 + C_2 \tag{11.3}$$

The second boundary condition depends on the heat transfer from the tip of the fin. If the fin may be assumed long and thin this is very small and may be assumed to be zero with very little error.

$$\therefore \qquad \left(\frac{\mathrm{d}\theta}{\mathrm{d}x}\right)_{x=l} = 0$$

$$\therefore \qquad mC_1 e^{ml} - mC_2 e^{-ml} = 0 \qquad (11.4)$$

Solution of equations (11.3) and (11.4) yields the values of C_1 and C_2, i.e.,

$$C_1 = \frac{\theta_0 e^{-ml}}{e^{ml} + e^{-ml}}, \quad \text{and} \quad C_2 = \frac{\theta_0 e^{ml}}{e^{ml} + e^{-ml}}$$

Substitution of these values back into equation (11.1) gives

$$\theta = \theta_0 \left[\frac{e^{m(l-x)} + e^{-m(l-x)}}{e^{ml} + e^{-ml}}\right]$$

$$\therefore \qquad \frac{\theta}{\theta_0} = \frac{\cosh m(l-x)}{\cosh ml} \qquad (11.5)$$

Even though it was assumed that $(\mathrm{d}\theta/\mathrm{d}x)_{(x=l)} = 0$, the temperature at the end of the fin is still above t_s, and is given by

$$\theta_l = \frac{\theta_0}{\cosh ml} \qquad (11.6)$$

This is obtained by putting $x = l$ in equation (11.5).

The total heat transfer from the fin is obtained by considering the conduction into the fin at the root. Thus:

$$Q_0 = -kA \left(\frac{\mathrm{d}\theta}{\mathrm{d}x}\right)_{x=0}$$

$$= mkA\theta_0 \left[\frac{\sinh m(l-x)}{\cosh ml}\right]_{x=0}$$

$$= mkA\theta_0 \tanh ml \qquad (11.7)$$

This result applies equally to the straight fin and spine, the appropriate value of m has merely to be substituted.

If the fin is comparatively short the assumption of no heat transfer from the tip of the fin is not valid. Under these conditions the heat transfer at the tip is given by

$$-kA\left(\frac{d\theta}{dx}\right)_{x=l} = +hA\theta_l$$

$$\therefore \quad -k(mC_1 e^{ml} - mC_2 e^{-ml}) = +h\theta_l \qquad (11.8)$$

The constants C_1 and C_2 may now be obtained by solving equations (11.3) and (11.8). Substituting for C_2 in (11.8) and eliminating θ_l by using (11.1):

$$-k[mC_1 e^{ml} - m(\theta_0 - C_1)e^{-ml}] = +h[C_1 e^{ml} + (\theta_0 - C_1)e^{-ml}]$$

This then gives:

$$C_1 = \frac{\theta_0[e^{-ml} - (h/km)e^{-ml}]}{(e^{ml} + e^{-ml}) + (h/km)(e^{ml} - e^{-ml})}$$

and

$$C_2 = \frac{\theta_0[e^{ml} + (h/km)e^{ml}]}{(e^{ml} + e^{-ml}) + (h/km)(e^{ml} - e^{-ml})}$$

and on substituting back into equation (11.1) gives

$$\frac{\theta}{\theta_0} = \frac{e^{m(l-x)} + e^{-m(l-x)} + (h/km)[e^{m(l-x)} - e^{-m(l-x)}]}{(e^{ml} + e^{-ml}) + (h/km)(e^{ml} - e^{-ml})}$$

which may be expressed as

$$\frac{\theta}{\theta_0} = \frac{\cosh m(l-x) + (h/km)\sinh m(l-x)}{\cosh ml + (h/km)\sinh ml} \qquad (11.9)$$

The temperature difference at the end of the fin is given by

$$\theta_l = \frac{\theta_0}{\cosh ml + (h/km)\sinh ml} \qquad (11.10)$$

The heat transfer from the fin is obtained as before by considering $(d\theta/dx)_{x=0}$. Thus

$$Q_0 = -kA\left(\frac{d\theta}{dx}\right)_{x=0}$$

$$= -kA\theta_0\left[\frac{-m\sinh m(l-x) - (h/k)\cosh m(l-x)}{\cosh ml + (h/km)\sinh ml}\right]_{x=0}$$

$$= mkA\theta_0\left[\frac{\sinh ml + (h/km)\cosh ml}{\cosh ml + (h/km)\sinh ml}\right]$$

$$= mkA\theta_0\left[\frac{\tanh ml + h/km}{1 + (h/km)\tanh ml}\right] \qquad (11.11)$$

EXAMPLE 11.1

A transistor heat sink is a 100 mm length of aluminium section as shown consisting of a 70 mm × 100 mm plate with 12 integral fins 25 mm high by 1 mm thick. If the plate is at 45 K above the surroundings find the percentage of heat transfer from the sink that occurs from the fins. $k = 0.15\,\mathrm{kW/(m\ K)}$, $h = 0.03\,\mathrm{kW/(m^2\ K)}$. Neglect heat transfer from the plate and fin edges. (*The City University*).

Solution. The plate surface temperature excess is 45 K, so heat transfer by convection from the plate is $(40 + (4 \times 6)) \times 100 \times 2 \times 0.03 \times 45 \times 10^{-6} = 0.0173\,\mathrm{kW}$.

For the fins, $m = (2 \times 0.03 \times 1000/0.15 \times 1)^{0.5} = 20.0$

For fins 25 mm in height, $ml = 20.0 \times 25.0/1000 = 0.5$

Tanh $ml = 0.462$. Hence heat transfer from 12 fins 100 mm long

$= 20.0 \times 0.15 \times 1 \times 45 \times 0.462 \times 100 \times 12/10^6 = 0.075\,\mathrm{kW}$

The total heat transfer is $0.0923\,\mathrm{kW}$, 81.3 per cent of this being from the fins.

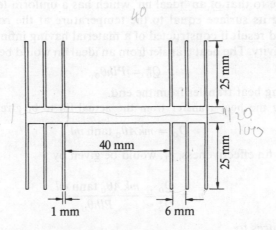

Fig. Example 11.1.

11.2 Limit of Usefulness of the Straight Fin

It is important to recognize the fact that fins may not necessarily improve heat transfer from a surface, and the conditions under which fins will not be useful must be investigated before any design work is contemplated. There are in fact three possibilities which arise from the particular value of the dimensionless grouping (h/km) which occurs in equation (11.11). By inspection:

$$\text{when } h/km > 1 \text{ then } Q_0 < h A \theta_0$$
$$\text{when } h/km = 1 \text{ then } Q_0 = h A \theta_0$$
$$\text{when } h/km < 1 \text{ then } Q_0 > h A \theta_0$$

Clearly, the requirement for increasing heat transfer from the surface is that $h/km < 1$, so that Q_0 is greater than $h A \theta_0$ which is the heat transfer from the surface having no fins. The possibility of using fins for insulation arises when $h/km > 1$, and $Q_0 < h A \theta_0$. It will be found that to achieve a value of $h/km > 1$, the resulting fins are so thick that the surface is virtually completely covered in insulation.

11.3 Fin and Finned Surface Effectiveness

A fin effectiveness may be defined by relating the actual fin performance to that of an 'ideal fin' which has a uniform temperature all along its surface equal to the temperature at the root. Such a fin would result if constructed of a material having infinite thermal conductivity. The heat transfer from an ideal fin would be defined by

$$Q_0^* = Plh\theta_0 \qquad (11.12)$$

neglecting heat transfer from the end.

Taking the heat transfer from the actual fin to be given by (11.7)

$$Q_0 = mkA\theta_0 \tanh ml$$

then the fin effectiveness, η_f, would be given by

$$\eta_f = \frac{Q_0}{Q_0^*} = \frac{mkA\theta_0 \tanh ml}{Plh\theta_0}$$

This reduces to

$$\frac{Q_0}{Q_0^*} = \frac{A^{\frac{1}{2}}k^{\frac{1}{2}} \tanh ml}{h^{\frac{1}{2}}P^{\frac{1}{2}}l}$$

$$\therefore \quad \frac{Q_0}{Q_0^*} = \frac{\tanh ml}{ml} \qquad (11.13)$$

If the fin which has a significant end heat transfer is compared with the ideal fin as defined by (11.12) then

$$\eta_f = \frac{\tanh ml + h/km}{ml + (hl/k)\tanh ml} \qquad (11.14)$$

The fin effectiveness is a useful idea in relation to the next topic to be considered, the overall heat transfer coefficients of surfaces which have fins. In Chapter 3, overall coefficient were derived for plane and cylindrical surfaces. Similar coefficients can be written for surfaces, both plane and cylindrical, on which fins have been added.

In the derivation of (11.13) it is seen that

$$(\eta_f Pl)h\theta_0 = mkA\theta_0 \tanh ml$$

so η_f may be interpreted as the fraction of fin area which may be regarded as being at θ_0 all over for purposes of calculating heat transfer. A function η_{fe} is now introduced which is the fraction of area of a *finned surface* at θ_0. If A_S and A_R are the total fin surface area and fin root area per unit area of primary or basic surface, respectively, then the total area of surface at θ_0 is $1 - A_R + \eta_f A_S$. As the total area is $1 - A_R + A_S$, the ratio of actual to ideal heat transfer from a finned surface is

$$\eta_{fe} = \frac{(1 - A_R + \eta_f A_S)h\theta_0}{(1 - A_R + A_S)h\theta_0} = \frac{1 - A_R + \eta_f A_S}{1 - A_R + A_S} \qquad (11.15)$$

An alternative method of assessing a finned surface is to compare its performance with that of the surface without fins, thus a surface 'coefficient of performance' would be given by

$$\text{C.O.P.} = \frac{(1 - A_R + \eta_f A_S)h\theta_0}{1 \times h\theta_0} = 1 - A_R + \eta_f A_S \qquad (11.16)$$

11.4 Overall Coefficients of Finned Surfaces

Fins are often added to only one surface to reduce the thermal resistance on that side. However, Fig. 11.4 shows a plane surface

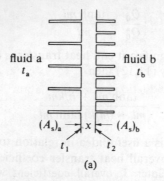

(a)

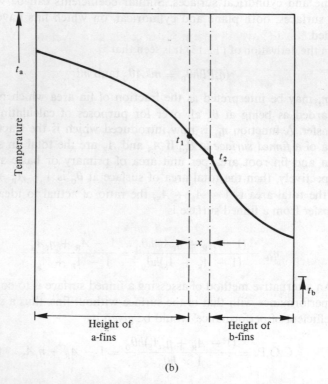

(b)

Fig. 11.4. (a) Plane finned surfaces separating two fluids. (b) Temperature pro-
files for (a). Heat transfer is to the a-fins from the fluid at t_a and from the b-fins
to the fluid at t_b.

with fins both sides. Heat transfer to the a-fins per unit plane area is
given by:

$$Q = -h_a(1 - A_R + \eta_f A_S)_a (t_1 - t_a)$$

Similarly, from the b-fins to fluid b

$$Q = -h_b(1 - A_R + \eta_f A_S)_b (t_b - t_2)$$

The heat transfer by conduction across the slab is

$$Q = -\frac{k}{x}(t_2 - t_1)$$

These three quantities are equal and combine to give

$$Q = -U(t_b - t_a)$$

where U is the overall heat transfer coefficient, given by

$$U = 1 \bigg/ \left\{ \frac{1}{h_a(1 - A_R + \eta_f A_S)_a} + \frac{x}{k} + \frac{1}{h_b(1 - A_R + \eta_f A_S)_b} \right\} \quad (11.17)$$

In this analysis, the group $(1 - A_R + \eta_f A_S)_a$ means that A_R, η_f and A_S all refer to the a-fins, and similarly for the b-fins.

A similar result may be obtained for a tube finned internally and externally, as shown in Fig. 11.5. Unit length of tube may be considered and the following three equations for heat transfer may be written:

Convection inside: $\qquad Q = -h_a 2\pi r_1 (1 - A_R + \eta_f A_S)_a (t_1 - t_a)$

Conduction: $\qquad Q = -\dfrac{2\pi k}{\ln r_2/r_1}(t_2 - t_1)$

Convection outside: $\qquad Q = -h_b 2\pi r_2 (1 - A_R + \eta_f A_S)_b (t_b - t_2)$

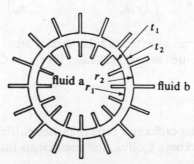

Fig. 11.5. *Cylindrical finned surfaces separating two fluids.*

These equations then lead to the result

$$Q = -U_L(t_b - t_a),$$

where

$$U_L = 1 \bigg/ \bigg\{ \frac{1}{2\pi r_1(1 - A_R + \eta_f A_S)_a h_a} + \frac{\ln r_2/r_1}{2\pi k}$$

$$+ \frac{1}{2\pi r_2(1 - A_R + \eta_f A_S)_b h_b} \bigg\} \qquad (11.18)$$

This has units of kW/(m K) or the equivalent. It is to be noted that $(1 - A_R + \eta_f A_S)_a$ is the fully effective surface area per unit area of the a-primary surface. This must then be multiplied by the area of primary surface per unit length, $2\pi r_1$. (Suffix L denotes U for unit length of tube).

Alternative expressions for U in terms of unit area of tube surface may also be obtained. If U_L in equation (11.18) is divided by $2\pi r_2$, the area of external primary surface per unit length, the result is

$$U_A = 1 \bigg/ \bigg\{ \frac{r_2}{r_1(1 - A_R + \eta_f A_S)_a h_a} + \frac{r_2 \ln r_2/r_1}{k}$$

$$+ \frac{1}{(1 - A_R + \eta_f A_S)_b h_b} \bigg\} \qquad (11.19)$$

This has units of kW/(m^{2}K) where the area refers to the external tube surface (primary) area. The corresponding result for the unfinned surface is

$$U_A = 1 \bigg/ \bigg(\frac{r_2}{r_1 h_a} + \frac{r_2 \ln r_2/r_1}{k} + \frac{1}{h_b} \bigg) \qquad (11.20)$$

This is again in terms of external tube surface area.

These results are used in heat exchanger theory, in Chapter 12.

EXAMPLE 11.2

A stainless steel heat exchanger tube is 25 mm outside diameter with a 2·5 mm wall thickness. Convection coefficients inside and outside are 6·0 and 1·0 kW/(m² K) and the thermal conductivity is 0·04 kW/(m K). A similar tube has 20 axial fins 14 mm high by 2 mm thick. Find the overall coefficient in both cases, and in each case state which thermal resistance is controlling.

Solution. In the first case, equation (11.18) simplified for no fins becomes equation (3.17) with only one conduction term.

$$\therefore U_L = 1 \Big/ \left\{ \frac{1}{2\pi \times 0{\cdot}01 \times 6{\cdot}0} + \frac{\ln 1{\cdot}25}{2\pi \times 0{\cdot}04} + \frac{1}{2\pi \times 0{\cdot}0125 \times 1{\cdot}0} \right\}$$

$$= 1/(2{\cdot}65 + 0{\cdot}887 + 12{\cdot}75)$$

$$= 1/16{\cdot}3 = 0{\cdot}0614 \text{ kW/(m K)}$$

The convection resistance on the outside is clearly the largest and is therefore controlling, meaning that to reduce the overall resistance greatest benefit will be obtained by reducing this part of it.

In the second case, fins are added to the outside surface. The fin efficiency, $\eta = (\tanh ml)/ml$. Considering 1 m length,

$$m = \sqrt{(hP/ka)} = \sqrt{(1{\cdot}0 \times 2{\cdot}0/0{\cdot}04 \times 0{\cdot}002)}$$

$$= 158 \text{ and } ml = 2{\cdot}22$$

$$\eta = (\tanh 2{\cdot}22)/2{\cdot}22 = 0{\cdot}977/2{\cdot}22 = 0{\cdot}44$$

For the finned surface, $2\pi r_0 = 2\pi \times 0{\cdot}0125 = 0{\cdot}0785 \text{ m}^2/\text{m}$

$2\pi r_0 A_R = \text{root area/m length} = 20 \times 0{\cdot}002 \times 1 = 0{\cdot}04 \text{ m}^2/\text{m}$

$2\pi r_0 \eta A_S = \text{effective fin area/m length} = 20 \times 2 \times 0{\cdot}014 \times 0{\cdot}44$

$$= 0{\cdot}246 \text{ m}^2/\text{m}$$

$1/2\pi r_0 (1 - A_R + \eta A_S) h_0 = 1/(0{\cdot}0785 - 0{\cdot}04 + 0{\cdot}246) \times 1{\cdot}0$

$$= 3{\cdot}51$$

Equation (11.18) now gives

$$U_L = 1/(2{\cdot}65 + 0{\cdot}887 + 3{\cdot}51)$$

$$= 0{\cdot}142 \text{ kW/(m K)}$$

Although the inside and outside resistances are now similar, the outside one is just still controlling.

11.5 Numerical Relationships for Fins

The range of fin problems that may be analysed is greatly increased by the introduction of simple numerical relationships. Thus it is possible to include a variable convection coefficient, or even a

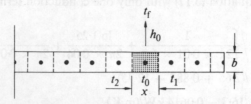

Fig. 11.6. Straight sided fin with h dependent on position.

transient analysis of a straight sided fin. Further, flat rectangular fins of the type encountered in air-conditioning equipment, can readily be analysed. The necessary relationships are deduced by the methods outlined in Chapters 4 and 5. Some examples are now given.

In Fig. 11.6 is shown a straight sided fin having a convection coefficient dependent on position. Separate relationships would be required for the root node, the tip node, and a further one for nodes in between. For central nodes:

$$(b \times 1)k\frac{(t_2 - t_0)}{x} + (b \times 1)k\frac{(t_1 - t_0)}{x} + h_0(2x \times 1)(t_f - t_0) = 0$$

$$\therefore \quad t_1 + t_2 + \left(\frac{2h_0x}{bk}\right)t_f - \left(2 + \frac{2h_0x}{bk}\right)t_0 = 0$$

In BASIC the statement would be:

$$X = (T(I-1) + T(I+1) + BTF(I)*TF)/(2{\cdot}0 + BTF(I)) \quad (11.21)$$

Used in an iterative procedure X is the new value of T(I), and

$$BTF(I) = 2{\cdot}0*H(I)*X/(B*TK)$$

where H(I) is the value of the convection coefficient at node I, $X = x$. $B = b$, $TK = k$, and $TF = t_f$. Corresponding BASIC statements for root and tip nodes are:

$$X = T(1) \quad\quad\quad\quad\quad\quad\quad\quad\quad\quad (11.22)$$

$$X = (T(I-1) + 0{\cdot}5*BTF(N)*TF)/(1{\cdot}0 + 0{\cdot}5*BTF(N)) \quad (11.23)$$

Hence it is seen that the root node is at the surface temperature T(1) and BTF(N) refers to the end node at I = N.

Fig. 11.7 shows the layout of a flat rectangular fin having a circular or elliptical root, which is approximated to the rectangular grid. Such a fin is usually symmetrical, so only one quarter need be considered. The general BASIC program in Chapter 4 is suitable for a

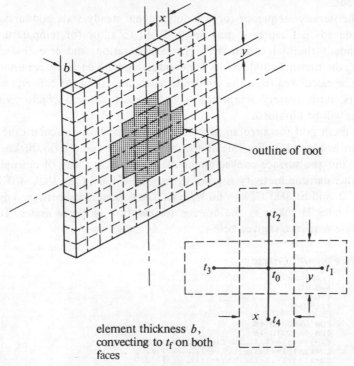

element thickness b, convecting to t_f on both faces

outline of root

Fig. 11.7. Flat rectangular fin with circular root.

steady state analysis of temperatures in such a fin. The equation for a field node is obtained from:

$$ kyb\left(\frac{t_1 - t_0}{x} + \frac{t_3 - t_0}{x}\right) + kxb\left(\frac{t_2 - t_0}{y} + \frac{t_4 - t_0}{y}\right) $$
$$ + 2hxy(t_f - t_0) = 0 \qquad (11.24) $$

Other equations are needed for side and corner boundary nodes, obtained from similar energy balances.

Other aspects of fin performance may be readily studied using programs in BASIC. For example, a fin root which is at the highest temperature in the fin also carried the greatest conduction heat load. Consequently the possibility arises of improving fin performance by using a material having temperature-dependent thermal conductivity, so that the highest thermal conductivity occurs where it is needed, i.e., at the root.

The iterative sequence for two-dimensional steady state conduction introduced in Chapter 4 may be modified to allow for temperature-dependent thermal conductivity. In each iteration, and at each field point, the thermal conductivity for conduction with each adjacent node must be calculated from the temperatures then existing. As convergence occurs, both correct temperatures and correct thermal conductivity values will be obtained.

A mesh grid for a rectangular plate fin mounted on a circular tube is shown in Fig. 11.7. For this fin the mesh size is A and the fin thickness is B, and the surface convection coefficient is H. By way of example, the thermal conductivity is given by $k = 50 \cdot 0 + 0 \cdot 1T$, so that at $0°C$, $k = 50$, and at $100°C$, $k = 60$ W/m K. The conductivity between adjacent nodes is taken at the average temperature of those nodes. The iterative sequence is given below.

BASIC Program Listing

```
500      CK=H*A*A/B
502      X=50. 0
503      Y=0. 1
510      ITER%=0
520      L%=0
530      FOR J%=2 TO 15
540      FOR I%=2 TO 16
550      C1=X1+(Y1*(T(I%,J%-1)+T(I%,J%)))/2. 0
560      C2=X1+(Y1*(T(I%-1,J%)+T(I%,J%)))/2. 0
570      C3=X1+(Y1*(T(I%,J%+1)+T(I%,J%)))/2. 0
580      C4=X1+(Y1*(T(I%+1,J%)+T(I%,J%)))/2. 0
590      C11=C1*(T(I%,J%-1))
600      C22=C2*(T(I%-1,J%))
610      C33=C3*(T(I%,J%+1))
620      C44=C4*(T(I%+1,J%))
630      K%=M%(I%,J%)
640      IF K%=1, GOTO 650
641      IF K%=2, GOTO 670
642      IF K%=3, GOTO 690
643      IF K%=4, GOTO 710
644      IF K%=5, GOTO 730
645      IF K%=6, GOTO 750
646      IF K%=7, GOTO 770
647      IF K%=8, GOTO 790
648      IF K%=9, GOTO 810
649      IF K%=10, GOTO 830
650      X=TIN
660      GOTO 840
670      X=(C11+2. 0*C44+C33+2. 0*CK*TCON)/(C1+C3+2. 0*(C4+CK))
```

```
680     GOTO 840
690     X=(C11+C44+CK*TCON)/(C1+C4+CK)
700     GOTO 840
710     X=(C11+(C44+C22)/2.0+CK*TCON)/(C1+(C2+C4)/2.0+CK)
720     GOTO 840
730     X=(C11+C22+CK*TCON)/(C1+C2+CK)
740     GOTO 840
750     X=((C11+C33)/2.0+C22+CK*TCON)/((C1+C3)/2.0+C2+CK)
760     GOTO 840
770     X=(C22+C33+CK*TCON)/(C2+C3+CK)
780     GOTO 840
790     X=(C22+2.0*C33+C44+2.0*CK*TCON)/(C2+C4+2.0*(C3+CK))
800     GOTO 840
810     X=(C11+C22+C33+C44+2.0*CK*TCON)/(C1+C2+C3+C4+(2.0*CK))
820     GOTO 840
830     X=0.0
840     DT=ABS(T(I%,J%)-X)
850     IF(DT>0.005)GOTO 870
860     L%=L%+1
870     T(I%,J%)=T(I%,J%)+1.9*(X-T(I%,J%))
880     NEXT I%
890     NEXT J%
900     ITER%=ITER%+1
910     IF(L%<210)GOTO 520
920     IF((210-L%)>0)GOTO 520
```

Problems

1. The diagram shows the cross-section of a nuclear reactor fuel element consisting of a uranium fuel rod 28 mm diameter contained in a magnox can which has longitudinal finning on its external surface. The fuel rod/can interface temperature is 430°C and the heat release rate is 65·6 kW per m length. Calculate the maximum temperature within the fuel rod, the tem-

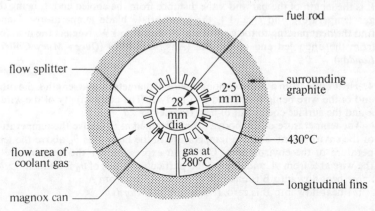

perature drop across the 2·5 mm wall of magnox, and the total surface area of longitudinal finning per unit length, given that the fin root is 40 per cent of the external can surface area, fin efficiency is 90 per cent (based on fins + splitters which also act as fins), mean coolant gas temperature 280°C, surface convection coefficient 0·8 kW/(m² K), k for magnox 0·15 kW/(m K), k for uranium fuel 0·0325 kW/(m K). (Ans.: 591°C, 11·5°C, 0·589 m²/m.) (*The City University*).

2. Write a program in BASIC (or FORTRAN if preferred) which will calculate the heat transfer rate from a 100 mm square fin, of thickness 0·5 mm, having a root approximating to a 25 mm diameter circle. Use a temperature-dependent thermal conductivity for the fin material, and investigate the percentage increase in fin heat transfer for a material which shows (a) 50 per cent and (b) 100 per cent increase in thermal conductivity over the temperature range between the root and surroundings temperature. In addition, investigate the effect of surface convection coefficient, for values typically in the range of (i) natural convection and (ii) forced convection.

3. A bar simulating a gas turbine blade, cooled at the root, is 10·2 cm long and has a cross-sectional area (A) 1·93 cm^2, and a perimeter (p) of 7·6 cm. Gas at 815°C streams across it, and one end is cooled to 483°C. The mean heat transfer coefficient for the gas flow conditions can be assumed constant over the surface at 0·284 kW/(m^{2}K), and the thermal conductivity of the material of the bar (k) is 26×10^{-3} kW/(m K). Show that

$$\frac{\theta_x}{\theta_r} = \frac{\cosh mL(1 - x/L)}{\cosh mL}$$

where

$$\theta_x = t_g - t_x, \quad \theta_r = t_g - t_r, \quad m = \sqrt{\frac{hp}{kA}}$$

L is the length of the bar and x the distance from the cooled end, t_g being the gas temperature and t_x and t_r the appropriate blade temperatures. Hence find the heat passing to the cooled end of the bar in kW. Neglect the heat lost from the uncooled end of the bar. (Ans.: 0·11 kW.) (*Queen Mary College, London*).

4. Heat flows from a body A along a wire of diameter d and length l, the other end of the wire being connected to a body B. The conductivity of the wire is k and the surface coefficient of heat transfer $\frac{1}{4}\alpha^2 kd$.

The temperature of the body A is maintained at θ_A above the temperature of the environment, and the temperature of the body B is θ_B above the temperature of the environment. Derive an expression for the temperature of the wire at x from A, and deduce the particular values of θ_B for which

(a) heat flow into B is one-half of the heat flow from A,
(b) heat flow into B is zero. (*University of Oxford*).

5. The cooling system of an electronic package has to dissipate 0·153 kW from the surface of an aluminium plate 100 mm × 150 mm. It is proposed to use 8 fins each 150 mm long and 1 mm thick. The temperature difference between the plate and surroundings is 50 K, the thermal conductivity of plate and fins is 0·15 kW/(m K), the convection coefficient is 0·04 kW/(m^2 K). Calculate the height of fin required and the effectiveness of the whole cooling surface. (Ans.: 30·3 mm, 88·4%.) (*The City University*).

6. A vertical pipe is 3 m long by 50 mm diameter. The surface is at 70°C and it convects to the surroundings at 15°C. Twelve rectangular fins are integral with the surface of the pipe; they are 3 m long by 40 mm in height (i.e., they extend 40 mm radially from the pipe surface and run the entire length of the pipe), and they are 2 mm thick. Show that natural convection over the pipe and fin surface is turbulent and that the heat transfer coefficient is to be determined from $Nu = 0\cdot129$ $(Gr\,Pr)^{0\cdot333}$ (where $\theta = 55°C$).

Determine: (a) the heat transfer coefficient, (b) the fin efficiency and (c) the rate of convection from the total pipe and fin surface. (Ans. (a) $5\cdot526$ W/m K, (b) $0\cdot926$, (c) $931\cdot9$ W.)

7. A shell and tube heat exchanger is to use copper tubes of outer diameter 15 mm having a wall thickness of 1 mm. The outer surface will have integral copper fins 1 mm thick and 10 mm in height, running axially along the tube. The fluid convection coefficient inside the tube is 480 W/m² K, and outside the tube it is 100 W/m² K.

Determine to the nearest whole number, the number of fins required to give an equal thermal resistance on both surfaces of the tube, and calculate the overall heat transfer coefficient per unit length of the tube. Take k for copper as 368 W/m K. (Ans. 7 fins, $(6\cdot364)$, $8\cdot384$ W/m K.)

REFERENCES

1. Jakob, M. *Heat Transfer*, Vol. 1, John Wiley and Sons, Inc., New York (1949).
2. Eckert, E. R. G., and Drake, R. M. *Analysis of Heat and Mass Transfer*, McGraw-Hill Book Company, Inc., New York (1972).
3. Chapman, A. J. *Heat Transfer*, 3rd ed., The Macmillan Company, New York (1974).

12

Heat exchangers

Much of the basic conduction and convection theory finds its greatest application in the heat exchanger. Whenever it is necessary to transfer energy from one fluid to another in large quantities, some form of heat exchanger is used. The most common form of heat exchanger is that in which two fluid streams pass through in steady flow, and heat transfer takes place through a separating wall. Mechanisms involved are therefore convection to or from the solid surface and conduction through the wall. The wall may be corrugated or finned to increase turbulence and the heat transfer area.

The thermal capacity of a heat exchanger is usually kept small, and is of significance only in transient conditions. However, a regenerative type of heat exchanger does have a large thermal capacity matrix through which the hot and cold fluids pass alternately. By this means energy is transferred indirectly between the fluids.

This chapter is concerned only with non-regenerative heat exchangers in which the fluids are separated. Other types of heat exchanger exist in which the fluids mix. These include cooling towers and jet condensers, for example. The basic principles will be considered in relation to the simplest types only.

12.1 Types of Heat Exchanger, and Definitions

The two basic types of heat exchanger are the in-line or uni-directional flow exchanger and the cross-flow exchanger. Flow is along the same axis in the in-line exchanger, but the two fluids may flow in the same or opposite directions giving rise to the names parallel and counter flow. The in-line exchanger may consist simply of two concentric tubes, one fluid flowing in the inner tube and the other in the annulus. Alternatively, there may be a number of tubes within a large tube or shell and to increase heat transfer the shell fluid is made to flow partly across the tubes by means of baffles.

176

Counter and parallel flow also occur in plate heat exchanges in which the fluids flow between closely spaced plates sealed at the edges. Fig. 12.1 shows some simple in-line arrangements and Fig. 12.2 shows a part section of a shell and tube heat exchanger with baffles in the shell of the segmented and 'doughnut' type.

The cross-flow exchanger is, as its name implies, one in which the two fluid streams flow at right angles. Gas-to-gas heat exchangers

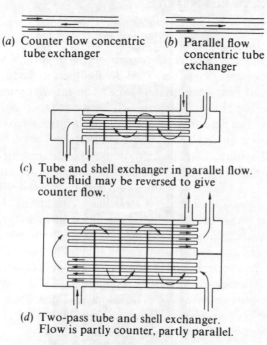

(a) Counter flow concentric tube exchanger

(b) Parallel flow concentric tube exchanger

(c) Tube and shell exchanger in parallel flow. Tube fluid may be reversed to give counter flow.

(d) Two-pass tube and shell exchanger. Flow is partly counter, partly parallel.

Fig. 12.1. Basic in-line heat exchangers

are often of this type. Their analysis is complicated because fluid temperatures vary in both the direction of flow and at right angles to that direction.

The temperature variations of the fluids in parallel and counter flow are shown in Fig. 12.3. Temperatures are plotted against length or area of heat exchanger surface. The inlet end, where length or area is zero is regarded as being the end where the hotter of the two fluids enters. The fluids are regarded as being hot or cold, for convenience, and t_h is a temperature of the hot fluid, t_c a temperature

Fig. 12.2. *A liquid/liquid shell and tube heat exchanger. This type of unit is used for cooling transformer oil, with water as the cooling medium. Pressure drops: oil flow in the shell, 5-12 psi, water flow in the tubes, 1-5 psi. The heat transfer area is in the range 110-1090 ft^2, and the heat transfer rate is in the range 70-1950 kW. Photograph by courtesy of Associated Electrical Industries Limited.*

of the cold fluid. Suffixes 1 and 2 are used for inlet and outlet of individual streams, and θ_i is the temperature difference between fluids at the inlet end and θ_o the difference at the outlet end of the exchanger. An important term in heat exchanger theory is the *capacity ratio C*. It is a ratio of the products of mass flow rate and

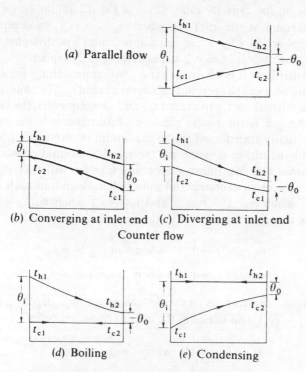

Fig. 12.3. *Temperature distributions.*

specific heat of each stream. It is always the ratio of the smaller product to the larger, since they are not necessarily equal. Thus, if $m_h c_{ph}$ is the 'capacity' of the hot stream and $m_c c_{pc}$ is that of the cold stream, then

if $m_h c_{ph} > m_c c_{pc}$,

$$C = \frac{m_c c_{pc}}{m_h c_{ph}} \tag{12.1}$$

and if $m_h c_{ph} < m_c c_{pc}$,

$$C = \frac{m_h c_{ph}}{m_c c_{pc}} \qquad (12.2)$$

In counter flow, the temperature distributions are rather different depending on the capacity ratio. Thus, in Fig. 12.3b, the temperatures are converging at the inlet end when $m_h c_{ph} > m_c c_{pc}$ and equation (12.1) applies. In Fig. 12.3c the temperatures are diverging at the inlet end when $m_c c_{pc} > m_h c_{ph}$ and equation (12.2) applies.

In parallel flow it is obvious that t_{c2} will approach t_{h2} for an infinitely long heat exchanger, but can never exceed t_{h2}. In counter flow it is quite normal for t_{c2} to exceed t_{h2} and, consequently, the counter flow exchanger is the more 'effective'. *Effectiveness* is the ratio of energy actually transferred to the maximum theoretically possible. Again, the definition depends on the relative thermal capacities of the streams. The maximum theoretical transfer will take place in counter flow in an exchanger of infinite length and, in such a case, $t_{c2} \to t_{h1}$ when $m_h c_{ph} > m_c c_{pc}$, and $t_{h2} \to t_{c1}$ when $m_h c_{ph} < m_c c_{pc}$. Thus the maximum transfers in the two cases are:

$$m_c c_{pc}(t_{h1} - t_{c1}) \quad \text{when} \quad m_h c_{ph} > m_c c_{pc}$$

$$m_h c_{ph}(t_{h1} - t_{c1}) \quad \text{when} \quad m_h c_{ph} < m_c c_{pc}$$

The actual transfers in the two cases are $m_c c_{pc}(t_{c2} - t_{c1})$ and $m_h c_{ph}(t_{h1} - t_{h2})$, and hence E, the effectiveness, becomes

$$E = \frac{t_{c2} - t_{c1}}{t_{h1} - t_{c1}} \quad \text{when} \quad m_h c_{ph} > m_c c_{pc} \qquad (12.3)$$

and

$$E = \frac{t_{h1} - t_{h2}}{t_{h1} - t_{c1}} \quad \text{when} \quad m_h c_{ph} < m_c c_{pc} \qquad (12.4)$$

These definitions may be used in either counter or parallel flow, but the value of E will be lower in parallel flow.

Temperature distributions with a change of phase are also shown in Fig. 12.3. These will occur in boiling, Fig. 12.3d, and condensing, Fig. 12.3e. Only the phase change takes place in the exchanger, so the temperature of the boiling or condensing fluid does not change. The temperature distributions are the same for both parallel and counter flow. The capacity ratio C becomes 0

for both boiling and condensing since the larger thermal capacity
is in each case infinite. This follows, since by definition, $c_p = dh/dt$
$= \infty$ when $dt = 0$. Equations (12.3) and (12.4) may be used in
condensing and boiling, respectively.

The other limit of capacity ratio is $C = 1$ and occurs when the
thermal capacities of the two streams are equal. This is not illustrated,
but it results in the temperature distributions being parallel straight
lines in the case of counter flow, θ being a constant over the whole
heat exchange area.

12.2 Determination of Heat Exchanger Performance

The primary purpose of a heat exchanger is to achieve the required
transfer rate using the smallest possible transfer area and fluid
pressure drop. A large exchanger can mean unnecessary capital
outlay and high pressure drop means a reduced efficiency of the
plant considered overall. Generally, a smaller exchanger can be
produced by finning surfaces to increase the overall heat transfer
coefficient. However, this leads to a higher fluid pressure drop,
and the best design is often a compromise between conflicting
requirements. In fact, a number of different designs for a given duty
may be acceptable.

The heat transfer requirement, Q, can be expressed in three ways:

$$Q = U_A A \theta_m = U_L L \theta_m \tag{12.5}$$

$$Q = m_c c_{pc}(t_{c2} - t_{c1}) \tag{12.6}$$

$$Q = m_h c_{ph}(t_{h1} - t_{h2}) \tag{12.7}$$

θ_m is a mean temperature difference between the fluids, and U_A
and U_L are mean coefficients, in $kW/(m^2 K)$ and $kW/(mK)$ or equivalent units, applicable over the entire area A or length L of the exchanger. It is general practice to work in terms of the external
surface area of the tubes in heat exchanger design, and the overall
coefficient U_A in terms of this area is given by equations (11.19)
for finned surfaces and (11.20) for plain surfaces.

12.2.1 Counter and Parallel Flow
If the mass flow rates and inlet and outlet temperatures are known, the
heat transfer Q will be known, but further details of the exchanger
cannot be specified until θ_m is known. θ_m can be derived as follows:

Consider an incremental area of heat exchanger surface as shown for either counter or parallel flow in Fig. 12.4. The heat transfer over the area dA can be expressed in three ways as before, thus

$$dQ = U_A \, dA\theta \tag{12.8}$$

$$dQ = m_c c_{pc} \, dt_c \tag{12.9}$$

$$dQ = m_h c_{ph} \, dt_h \tag{12.10}$$

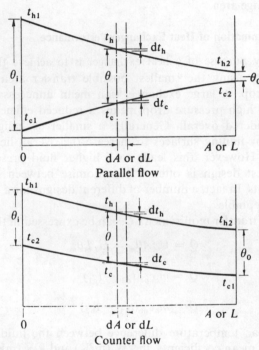

Fig. 12.4. *For the determination of logarithmic mean temperature difference*

The temperature difference at the point in question is

$$\theta = t_h - t_c$$

and the increment in temperature difference is

$$d\theta = d(t_h - t_c)$$
$$= dt_h - dt_c \tag{12.11}$$

If $d\theta$ is expressed in terms of dQ from (12.9) and (12.10),

$$d\theta = -\frac{dQ}{m_h c_{ph}} - \frac{dQ}{m_c c_{pc}} \quad \text{for parallel flow} \quad (12.12)$$

and

$$d\theta = -\frac{dQ}{m_h c_{ph}} + \frac{dQ}{m_c c_{pc}} \quad \text{for counter flow} \quad (12.13)$$

since dt_h is negative in both cases, and dt_c is positive for parallel flow and negative for counter flow. It follows that $d\theta/dQ$ has constant but different values for parallel and counter flow, and therefore

$$\frac{\theta_o - \theta_i}{Q} = -\left(\frac{1}{m_h c_{ph}} \pm \frac{1}{m_c c_{pc}}\right) \quad (12.14)$$

with $+$ for parallel flow and $-$ for counter flow. Further, dQ from equation (12.8) may be substituted in (12.12) and (12.13) to give

$$\frac{d\theta}{\theta} = -\left(\frac{1}{m_h c_{ph}} \pm \frac{1}{m_c c_{pc}}\right) U \, dA$$

This is integrated from 0 to A to give

$$\ln \frac{\theta_o}{\theta_i} = -\left(\frac{1}{m_h c_{ph}} \pm \frac{1}{m_c c_{pc}}\right) U_A A \quad (12.15)$$

The term in parentheses is now eliminated between (12.15) and (12.14) to give

$$Q = U_A A \frac{\theta_o - \theta_i}{\ln(\theta_o/\theta_i)} \quad (12.16)$$

This result is clearly identical in form to equation (12.5) and it is seen that

$$\theta_m = \frac{\theta_o - \theta_i}{\ln(\theta_o/\theta_i)} \quad (12.17)$$

This is the required logarithmic mean temperature difference. It is the same for counter and parallel flow, though θ_o and θ_i in terms of values of t_h and t_c are different as can be seen from Fig. 12.3.

EXAMPLE 12.1

0·2 kg/s of an alcohol is to be cooled from 75 to 35°C in a counter flow heat exchanger. Cooling water enters the exchanger at 12°C and at the rate of 0·16 kg/s. The convection coefficient between the alcohol and the tube wall is 0·34 kW/(m²K), and between the tube wall and the water, 0·225 kW/(m²K). The tubes may be assumed thin. c_p for the alcohol is 2·52 kJ/(kg K) and for water is 4·187 kJ/(kg K).

Calculate the capacity ratio, the effectiveness, and the area of the heat exchanger surface.

Solution. For the hot stream, alcohol,

$$m_h c_{ph} = 0.2 \times 2.52 = 0.504 \text{ kJ/(s K)}$$

For the cold stream, water,

$$m_c c_{pc} = 0.16 \times 4.187 = 0.671 \text{ kJ/(s K)}$$

From equation (12.2), $C = m_h c_{ph}/m_c c_{pc} = 0.504/0.671 = 0.75$. An energy balance gives

$$0.2 \times 2.52 \times (75 - 35) = 0.16 \times 4.187 \times (t_{c2} - 12)$$
$$20.15 = 0.671 t_{c2} - 8.05$$
$$\therefore \quad t_{c2} = 41.8°C$$

From equation (12.4),

$$E = \frac{t_{h1} - t_{h2}}{t_{h1} - t_{c1}} = \frac{75 - 35}{75 - 12} = 0.635$$

The heat exchange area may be found from equation (12.5). To find θ_m:

$$\theta_o = 35 - 12 = 23, \qquad \theta_i = 75 - 41.8 = 33.2$$
$$\therefore \quad \theta_m = \frac{23 - 33.2}{\ln(23/33.2)} = \frac{-10.2}{-\ln 1.44} = 28 \text{ K}$$

Since the tubes are thin, $r_1 = r_2$ in (11.20), so U_A is given by

$$\frac{1}{U_A} = \frac{1}{h_{\text{alcohol}}} + \frac{1}{h_{\text{water}}} = \frac{1}{0.34} + \frac{1}{0.225}$$

$$\therefore \qquad U_A = 0.1355 \text{ kW/(m}^2\text{K)}$$

Equation (13.5) gives $20.15 = U_A A \theta_m = 0.1355 \times A \times 28$

$$\therefore \qquad A = 5.31 \text{ m}^2$$

12.2.2 Cross Flow

Analysis of the cross-flow heat exchanger is more complicated owing to temperature variation across the flow. This variation will depend on whether the fluid is *mixed* or *unmixed*. A mixed fluid is free to move across the flow direction; an unmixed fluid is constrained in parallel flow passages. Thus, if an exchanger consisted of a bank of tubes placed across a duct, the fluid in the duct would be mixed while the fluid in the tubes would be unmixed.

Results of analyses of this type of exchanger are available as correction factors.[1,2] Equation (12.5) would become

$$Q = U_A A F \theta_m$$

where F is a factor to be obtained from the appropriate graph, and θ_m is the mean temperature difference (12.17), calculated for counter flow with the same inlet and outlet temperatures as for cross flow. Figure 12.5 shows F for a cross-flow exchanger with one fluid mixed and one fluid unmixed. In applying the factor F it does not matter whether the hotter fluid is mixed or unmixed.

12.3 Heat Exchanger Transfer Units

One would now expect to be able to go ahead and design a heat exchanger, using equations (12.5) to (12.7) and information from earlier chapters to evaluate U_A for the particular configuration in mind. However, U_A cannot be determined until something is known of the tube sizes and velocities of flow, and the method of procedure from theory so far developed can be extremely involved and iterative. For example, supposing the tube sizes, length and U_A were decided upon, in order to check the design performance the value of Q and outlet temperatures of the fluids must be regarded as unknowns and equations (12.5) to (12.7) cannot be solved directly

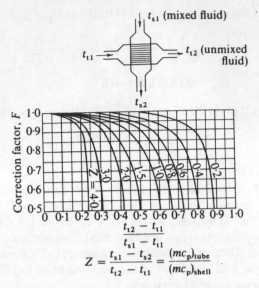

Fig. 12.5. Logarithmic temperature difference correction factor for cross flow, one fluid mixed, one fluid unmixed. From R. A. Bowman, A. E. Mueller, and W. M. Nagle, Trans. ASME, Vol. 62, p. 283 (1940). By permission of the American Society of Mechanical Engineers.

for Q, t_{c2}, and t_{h2}, because of the logarithmic form of θ_m. The approach using transfer units is very useful from this point of view. The method was developed by Kays and London.[3]

The effectiveness E, and capacity ratio C of a heat exchanger have already been defined. These quantities will now be used in conjunction with a new term, Number of Transfer Units, NTU, to determine heat exchanger performance. As with E and C, the definition of NTU depends on the relative magnitudes of the thermal capacities of the fluid stream. Thus,

$$NTU = \frac{U_A A}{m_c c_{pc}}, \qquad \text{when } m_h c_{ph} > m_c c_{pc} \qquad (12.18)$$

and,

$$NTU = \frac{U_A A}{m_h c_{ph}}, \qquad \text{when } m_h c_{ph} < m_c c_{pc} \qquad (12.19)$$

Thus the denominator is always the smaller thermal capacity. The performance of heat exchangers will now be examined using the definitions of C, E, and NTU in equations (12.1) to (12.4) and (12.18) and (12.19).

12.3.1 Counter Flow Exchanger

Let $m_h c_{ph}$ be assumed the smaller quantity, then the definitions of NTU, C, and E are

$$NTU = \frac{U_A A}{m_h c_{ph}}, \quad C = \frac{m_h c_{ph}}{m_c c_{pc}}, \quad E = \frac{t_{h1} - t_{h2}}{t_{h1} - t_{c1}}$$

Equations (12.9) and (12.10) for counter flow (where temperature increments are negative) give

$$-m_c c_{pc}\, dt_c = -m_h c_{ph}\, dt_h = dQ \qquad (12.20)$$

Now, $d\theta = d(t_h - t_c) = dt_h - dt_c$,

and $m_h c_{ph}(dt_h - dt_c) = dt_c(m_c c_{pc} - m_h c_{ph})$

using equation (12.20). Again, using (12.20), dt_c may be eliminated to give

$$m_h c_{ph}(dt_h - dt_c) = -\frac{dQ}{m_c c_{pc}}(m_c c_{pc} - m_h c_{ph})$$

$$= -dQ(1 - C)$$

Using equation (12.8) to eliminate dQ gives

$$dt_h - dt_c = -\frac{U_A\, dA\theta}{m_h c_{ph}}(1 - C)$$

$$\therefore \quad \frac{d\theta}{\theta} = -\frac{U_A\, dA}{m_h c_{ph}}(1 - C)$$

Integrating:

$$\ln\frac{\theta_o}{\theta_i} = \ln\frac{t_{h2} - t_{c1}}{t_{h1} - t_{c2}} = -\frac{U_A A}{m_h c_{ph}}(1 - C)$$

$$= -NTU(1 - C)$$

$$\therefore \quad \frac{t_{h2} - t_{c1}}{t_{h1} - t_{c2}} = e^{-NTU(1-C)}$$

The left-hand side of this equation may be manipulated as follows:

$$\frac{t_{h2} - t_{c1}}{t_{h1} - t_{c2}} = \frac{t_{h1} - t_{c1} - (t_{h1} - t_{h2})}{t_{h1} - t_{c1} - (t_{c2} - t_{c1})}$$

$$= \frac{t_{h1} - t_{c1} - (t_{h1} - t_{h2})}{t_{h1} - t_{c1} - C(t_{h1} - t_{h2})}, \quad \text{(using the definition of } C)$$

$$= \frac{1 - \dfrac{t_{h1} - t_{h2}}{t_{h1} - t_{c1}}}{1 - \dfrac{C(t_{h1} - t_{h2})}{(t_{h1} - t_{c1})}}$$

$$= \frac{1 - E}{1 - CE} = e^{-NTU(1-C)}$$

from the right-hand side, above. This final result is now rearranged to give

$$E = \frac{1 - e^{-NTU(1-C)}}{1 - C\,e^{-NTU(1-C)}} \tag{12.21}$$

If $m_c c_{pc}$ had been assumed the smaller quantity, the same equation would have been obtained, where E, NTU, and C would have then been defined by the alternative expressions.

A relationship exists, then, between E, NTU, and C given by equation (12.21). Using this result it is possible to determine outlet temperatures t_{c2} and t_{h2}, and Q, the overall heat transfer for a given design, without using a trial and error solution.

EXAMPLE 12.2

Determine the effectiveness and fluid outlet temperature of an oil cooler handling 0·5 kg/s of oil at an inlet temperature of 130°C. The mean specific heat is 2·22 kJ/(kg K). 0·3 kg/s of water entering at 15°C passes in counter flow at a rate of 0·3 kg/s. The heat transfer surface area is 2·4 m² and the overall heat transfer coefficient is known to be 1·53 kW/(m² K)

Solution. The thermal capacities are: oil, $0·5 \times 2·22 = 1·11$ kJ/(s K), water, $0·3 \times 4·182 = 1·255$ kJ/(s K)

$$\therefore \qquad C = 1·11/1·255 = 0·885$$

and,

$$NTU = \frac{1 \cdot 53 \times 2 \cdot 4}{1 \cdot 11} = 3 \cdot 31$$

Then,

$$E = \frac{t_{h1} - t_{h2}}{t_{h1} - t_{c1}} = \frac{1 - e^{-3 \cdot 31(1 - 0 \cdot 885)}}{1 - 0 \cdot 885 \, e^{-3 \cdot 31(1 - 0 \cdot 885)}}$$

$$= \frac{1 - e^{-0 \cdot 38}}{1 - 0 \cdot 885 \, e^{-0 \cdot 38}} = \frac{0 \cdot 316}{0 \cdot 395} = 0 \cdot 8$$

$$= \frac{130° - t_{h2}}{130° - 15°}$$

$$\therefore \qquad t_{h2} = 38 \cdot 0°C \qquad \text{(oil outlet)}$$

By enthalpy balance

$$(t_{c2} - t_{c1}) = \frac{1 \cdot 11 \times (130 - 38)}{1 \cdot 255} = 81 \cdot 5 \, \text{K}$$

$$\therefore \qquad t_{c2} = 96 \cdot 5°C \qquad \text{(water outlet)}$$

When U_A is not known, this must be determined from either equation (11.19) or (11.20), with the individual convection coefficients determined from the equation appropriate to the fluid, flow geometry and type of flow, as given in earlier chapters. It is convenient to use standard tube sizes to give a suitable value of Re and number of tubes for the specified mass flow. Several attempts may be necessary to achieve a suitable U_A combined with a fluid pressure loss which is acceptable.

12.3.2 Parallel Flow Exchanger
A similar analysis in parallel flow will yield the result

$$E = \frac{1 - e^{-NTU(1 + C)}}{1 + C} \tag{12.22}$$

Again this result is independent of which fluid stream has the smaller thermal capacity, provided the appropriate definitions of E, NTU, and C are used.

12.3.3 Limiting Values of C

It has already been noted that $C = 0$ in both condensing and boiling. When this is so both equation (12.21) and (12.22) reduce to

$$E = 1 - e^{-NTU} \tag{12.23}$$

Thus, the effectiveness is the same for both counter and parallel flow.

The other limiting value is $C = 1$ for equal thermal capacities and, in this case, for parallel flow equation (12.22) gives

$$E = \frac{1 - e^{-2NTU}}{2} \tag{12.24}$$

In the case of counter flow for $C = 1$ it is necessary to do a fresh analysis from first principles since equation (12.21) becomes indeterminate. For this case it is possible to write

$$E = (t_{h1} - t_{h2})/(t_{h1} - t_{c1})$$

and also

$$(t_{h1} - t_{h2}) = (t_{c2} - t_{c1})$$

Also

$$Q = U_A A(t_{h1} - t_{c2}) = mc_p(t_{h1} - t_{h2})$$

$$\therefore \quad (t_{h1} - t_{h2}) = NTU(t_{h1} - t_{c2})$$

E may be written as

$$E = \frac{t_{h1} - t_{h2}}{(t_{h1} - t_{h2}) - (t_{c1} - t_{h2})} = \frac{(t_{h1} - t_{c2})NTU}{(t_{h1} - t_{c2})NTU - (t_{c1} - t_{h2})}$$

But $(t_{c1} - t_{h2}) = -(t_{h1} - t_{c2})$

$$\therefore \quad E = \frac{NTU}{NTU + 1}, \quad \text{when } C = 1 \tag{12.25}$$

12.3.4 Cross-Flow Exchanger

Convenient graphical plots of effectiveness as a function of *NTU* and capacity ratio are available for cross flow. Figure 12.6 is for one fluid mixed and one fluid unmixed. When the capacity ratio of mixed to unmixed fluid is greater than 1, the NTU is then based on (mc_p) of the unmixed fluid.

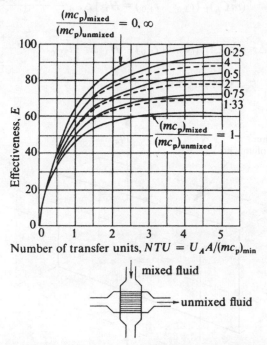

Fig. 12.6. Effectiveness vs. NTU for a cross-flow exchanger, one fluid mixed, one fluid unmixed. From Compact Heat Exchangers, by W. M. Kays and A. L. London, McGraw-Hill Book Company, Inc., New York (1958). Used by permission of McGraw-Hill Book Company.

12.4 Heat Exchange in Cross Flow

Cross-flow heat exchangers may be analysed most readily by numerical techniques, and Fig. 12.7 shows a typical model for consideration. For the nodal element, one fluid enters a control volume at t_1 behind the dividing wall at t_M, and the second fluid, flowing upwards, enters the control volume in front at t_2. For transient analysis, a second subscript is added to denote time, e.g., $t_{1,0}$, $t_{2,0}$, $t_{M,0}$. For the fluid flowing left to right, the following equations may be written:

Steady state: $(\dot{m}C_p)_1 \, (t_3 - t_1) = H \left[\dfrac{t_2 + t_4}{2} - \dfrac{t_1 + t_3}{2} \right]$ (12.26)

Transient: $(\dot{m}C_p)_1 \, (t_{3,0} - t_{1,0}) = H_1 \left[t_{M,0} - \dfrac{t_{1,0} + t_{3,0}}{2} \right]$ (12.27)

or: $\Delta\tau \, (\dot{m}C_p)_1 \, (t_{3,0} - t_{1,0}) = H_1 \, \Delta\tau \left[t_{M,0} - \dfrac{t_{1,0} + t_{3,0}}{2} \right]$

$\qquad\qquad - (mC_v)_1 (t_{3,1} - t_{3,0})$ (12.28)

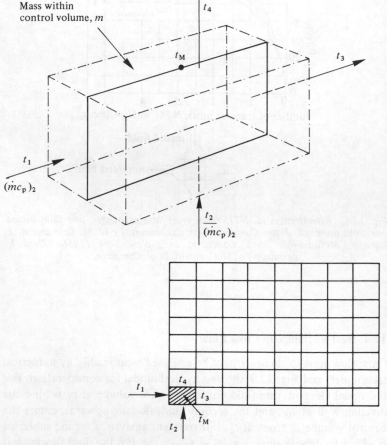

Mass within control volume, m

Fig. 12.7. Thermal model for cross-flow heat exchange.

The second transient equation is more accurate and complete. The additional term, which is the increase in stored energy in the control volume, is very small for gas to gas heat exchange, and is eliminated in equation (12.27). Equation (12.28) when re-arranged may be expressed as: increase of stored energy = heat flow across wall of control volume + enthalpy flow in − enthalpy flow out. In practice, equation (12.28) is not often used. In equations (12.27) and (12.28), $H_1 = h_1 A_1 =$ convection coefficient × area of wall, for fluid 1; and similarly for fluid 2, $H_2 = h_2 A_2$. In equation (12.26) where the wall temperature is not of importance, the overall thermal resistance between fluids is given by:

$$1/H_1 + 1/H_2 = (H_1 + H_2)/(H_1 H_2)$$

Hence the overall heat transfer coefficient is

$$H = H_1 H_2 / (H_1 + H_2)$$

Introducing the groups:

$$NTU_1 = H_1 / (\dot{m} C_p)_1 \text{ and } NTU_2 = H_2 / (\dot{m} C_p)_2$$

Equation (12.27) becomes, for the unknown $t_{3,0}$:

$$t_{3,0} = \left[\frac{2\,NTU_1}{2 + NTU_1} \right] t_{M,0} + \left[\frac{2 - NTU_1}{2 + NTU_1} \right] t_{1,0} \qquad (12.29)$$

A similar equation for $t_{4,0}$ would be:

$$t_{4,0} = \left[\frac{2\,NTU_2}{2 + NTU_2} \right] t_{M,0} + \left[\frac{2 - NTU_2}{2 + NTU_2} \right] t_{2,0} \qquad (12.30)$$

In order to calculate t_3 and t_4 after the next time step, an equation for a new t_M must be obtained. Thus:

$$\left[\frac{t_{1,0} + t_{3,0}}{2} - t_{M,0} \right] H_1 + \left[\frac{t_{2,0} + t_{4,0}}{2} - t_{M,0} \right] H_2$$
$$= \frac{(m C_p)_M}{\Delta \tau} (t_{M,1} - t_{M,0}) \qquad (12.31)$$

and hence

$$t_{M,1} = \frac{H_1 \Delta\tau}{2(mC_p)_M} \left[t_{1,0} + t_{3,0} \right] + \frac{H_2 \Delta\tau}{2(mC_p)_M} \left[t_{2,0} + t_{4,0} \right]$$

$$+ t_{M,0} \left[1 - \frac{(H_1 + H_2)\Delta\tau}{(mC_p)_M} \right] \qquad (12.32)$$

where $(mC_p)_M$ is the product of mass and specific heat of the metal element. For steady state, an enthalpy balance holds:

$$(\dot{m}C_p)_1 \, (t_3 - t_1) = (\dot{m}C_p)_2 \, (t_4 - t_2) \qquad (12.33)$$

From this equation, and equation (12.26), the following relations for t_3 and t_4 are obtained:

$$t_3 = \left[\frac{2\,NTU}{2 + NTU + C \times NTU} \right] t_2 + \left[1 - \frac{2\,NTU}{2 + NTU + C \times NTU} \right] t_1$$

$$(12.34)$$

$$t_4 = \left[\frac{2 \times C \times NTU}{2 + NTU + C \times NTU} \right] t_1 + \left[1 - \frac{2 \times C \times NTU}{2 + NTU + C \times NTU} \right] t_2$$

$$(12.35)$$

where $C = (\dot{m}C_p)_1/(\dot{m}C_p)_2$ and $NTU = H/(\dot{m}C_p)_1$

The above equations may be used in the analysis of transient and steady state behaviour of cross-flow heat exchangers. As shown in Fig. 12.7 outlet temperatures from one node become the inlet temperatures of the next node, so that the field of nodes may be solved in simple computer routines. Examples are given in Ref. 4. The equations to be used in any analysis are (12.29), (12.30), (12.32), (12.34), and (12.35). In each case the last term of the equation must remain positive for the solution to be stable. This sets a limit on the values of the variables occurring.

12.4.1 Rotary Regenerators

Rotary regenerators or thermal wheels are being used increasingly in energy conservation measures, and the steady state equations (12.26) and (12.33) above may be applied to their analysis. Figure 12.8 shows an element of gas flow in heat exchange with an element of matrix flow. Results corresponding to equations (12.34) and (12.35) are obtained:

$$t_2 = \left[\frac{2\,NTU_1}{2 + NTU_1 + C_1 \times NTU_1} \right] t_A + \left[1 - \frac{2\,NTU_1}{2 + NTU_1 + C_1 \times NTU_1} \right] t_1$$

(12.36)

$$t_B = \left[\frac{2 \times C_1 \times NTU_1}{2 + NTU_1 + C_1 \times NTU_1} \right] t_1 + \left[1 - \frac{2 \times C_1 \times NTU_1}{2 + NTU_1 + C_1 \times NTU_1} \right] t_A$$

(12.37)

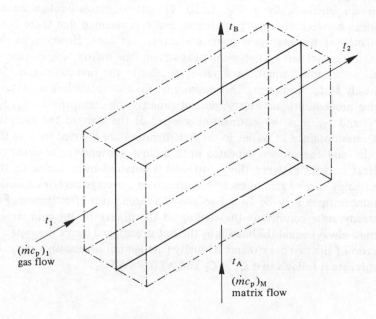

Fig. 12.8. Cross-flow heat exchange in a regenerator.

For the element shown, the gas stream entering at t_1 is in heat exchange with a moving solid matrix stream entering at t_A, and having a thermal capacity rate of $(\dot{m} C_p)_M$. In the two equations above:

$$NTU_1 = H_1/(\dot{m} C_p)_1 \quad \text{and} \quad C_1 = (\dot{m} C_p)_1/(\dot{m} C_p)_M$$

Later in the cycle, the matrix is in heat exchange with a second gas stream, and similar equations may be written, including the terms:

$$NTU_2 = H_2/(\dot{m}C_p)_2 \text{ and } C_2 = (\dot{m}C_p)_2/(\dot{m}C_p)_M$$

Usually

$$NTU_2 = NTU_1 \text{ and } C_2 = C_1$$

The elements of the form considered in Fig. 12.8 are pieced together as shown diagrammatically in Fig. 12.9. Imagine Fig. 12.9 as a fixed grid or series of control volumes with the fluids and matrix flowing through in the direction of the arrows. The actual thermal wheel is shown schematically in Fig. 12.10. The matrix passes through sealing zones between the two gas streams, and it is assumed that there is no change in the matrix temperature across this zone. However, as the matrix passes out of the second gas stream, the matrix temperature at all nodes must equal those values at inlet to the first gas stream, for steady state conditions. This means that for a correct solution an iterative procedure is necessary, since the matrix inlet temperature t_A, t_K, t_L and t_M must be assumed or guessed at the start of the analysis. Values obtained at outlet in the first iteration are inserted back in the inlet and the process repeated until a close convergence is obtained. Heat transfer between the gas streams is obtained by summing up the enthalpy loss or gain of each fluid stream, and average gas stream outlet temperatures may be found to give the regenerator effectiveness. For steady state conditions the cooling of the matrix by the cold stream must always equal the heating by the hot stream, and hence the capacity ratio of the two gas streams is usually (though not necessarily) unity. In this case it follows that $C_1 = C_2$ and $NTU_1 = NTU_2$.

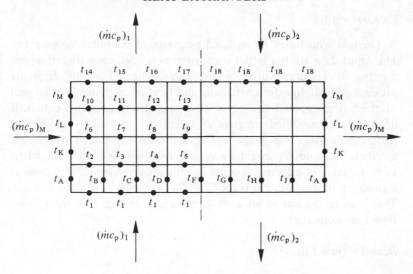

Fig. 12.9. Nodal system for the Thermal Wheel.

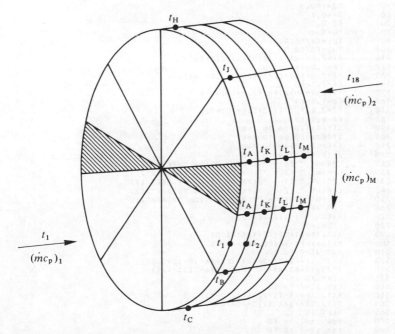

Fig. 12.10. Diagrammatic representation for the Thermal Wheel.

EXAMPLE 12.3

A Thermal Wheel may be analysed by means of the following program. The input data are the initial stream temperatures, capacity ratios, and number of Transfer Units for the whole wheel. For this, *H* is the product of heat transfer coefficient and heat transfer area for that part of the wheel exposed to one or other gas stream (assumed equal). If desired, in a modified program *H* could be calculated for different matrix hole sizes and hence surface areas for a specified wheel size. Similarly C_1 and C_2 could be varied, depending on different wheel sizes, rotating at different speeds. Results are given for Effectiveness as a function of Transfer Units for $C_1 = C_2$ for values of $0 \cdot 1$, $0 \cdot 5$ and $1 \cdot 0$. The form of the curves are seen to be similar to Fig. 12.6 for a cross-flow heat exchanger.

Basic Program Listing

```
20C       THERMAL WHEEL ANALYSIS
30        DIM A(10,10),T(10,10),AA(10,10),TT(10,10)
40        T1=500.0
50        T2=100.0
60        C1=0.1
70        C2=0.1
80        TN=5.0
90        TN1=TN/10.0
100       TN2=TN/10.0
110       F1=2.0*TN1*C1/(TN1*C1+2.0+TN1)
120       F2=1.0-F1
130       F3=2.0*TN2*C2/(TN2*C2+2.0+TN2)
140       F4=1.0-F3
150       F5=2.0*TN1/(TN1+2.0+TN1*C1)
160       F6=1.0-F5
170       F7=2.0*TN2/(TN2+2.0+TN2*C2)
180       F8=1.0-F7
190       IF(F1>1.0)GO TO 740
200       IF(F3>1.0)GO TO 740
210       IF(F5>1.0)GO TO 740
220       IF(F7>1.0)GO TO 740
225       LPRINT "  T1 = ";LPRINT USING "###.##";T1;
230       LPRINT "  T2 = ";LPRINT USING "###.##";T2;
235       LPRINT "  C1 = ";LPRINT USING "#.###";C1;
240       LPRINT "  C2 = ";LPRINT USING "#.###";C2;
250       LPRINT "  NTU = ";LPRINT USING "#.###";TU
260       FOR J%=1 TO 10
270       AA(J%,10)=T1;NEXT J%
280       ITER%=0
290       A(1,1)=F1*T1+F2*AA(10,10)
300       T(1,1)=F5*AA(10,10)+F6*T1
310       FOR J%=2 TO 10
320       A(J%,1)=F1*T(J%-1,1)+F2*AA(10-J%+1,10)
330       T(J%,1)=F5*AA(10-J%,10)+F6*T(J%-1,1)
335       NEXT J%
340       FOR I%=2 TO 10
350       A(1,I%)=F1*T1+F2*A(1,I%-1)
360       T(1,I%)=F5*A(1,I%-1)+F6*T1
365       NEXT I%
370       FOR J%=2 TO 10
380       FOR I%=2 TO 10
390       A(J%,I%)=F1*T(J%-1,I%)+F2*A(J%,I%-1)
```

```
400      T(J%,I%)=F5*A(J%,I%-1)+F6*T(J%-1,I%)
404      NEXT I%
405      NEXT J%
410      AA(1,1)=F3*T2+F4*A(10,10)
420      TT(1,1)=F7*A(10,10)+F8*T2
430      FOR J%=2 TO 10
440      AA(J%,1)=F3*TT(J%-1,1)+F4*A(10-J%+1,10)
450      TT(J%,1)=F7*A(10-J%+1,10)+F8*TT(J%-1,1)
455      NEXT J%
460      FOR I%=2 TO 10
470      AA(1,I%)=F3*T2+F4*AA(1,I%-1)
480      TT(1,I%)=F7*AA(1,I%-1)+F8*T2
485      NEXT I%
490      FOR J%=2 TO 10
500      FOR I%=2 TO 10
510      AA(J%,I%)=F3*TT(J%-1,I%)+F4*AA(J%,I%-1)
520      TT(J%,I%)=F7*AA(J%,I%-1)+F8*TT(J%-1,I%)
524      NEXT I%
525      NEXT J%
530      X=0
540      FOR J%=1,TO 10
550      X=X+AA(J%,10);NEXT J%
560      Z=ABS(X-Y)
570      IF(Z<0.001)GO TO 610
580      Y=X
590      ITER%=ITER%+1
600      GO TO 290
610      TOUT2=0.0
620      FOR I%=1 TO 10
630      TOUT2=TOUT2+TT(10,I%)/10.0
635      NEXT I%
640      EFF=(TOUT2-T2)/(T1-T2)
650      LPRINT "    Gas Stream Outlet Temperatures:"
660      LPRINT USING "####.##";T(10,I%) FOR I%=1 TO 10
670      LPRINT "    Air Stream Outlet Temperatures:"
680      LPRINT USING "####.##";TT(10,I%) FOR I%=1 TO 10
690      LPRINT "    Effectiveness = ";LPRINT USING"#.###";EFF;
700      LPRINT " for ";LPRINT USING"##.###";TU;LPRINT" Transfer Units"
710      LPRINT "Number of Iterations to Converge = ";
720      LPRINT USING "###";ITER
730      GO TO 790
740      LPRINT "    Check Data - Calculations Unstable:"
750      LPRINT "    F1 = ";LPRINT USING "##.##";F1;
760      LPRINT " F3 = ";LPRINT USING "##.##";F3;
770      LPRINT " F5 = ";LPRINT USING "##.##";F5;
780      LPRINT " F7 = ";LPRINT USING "##.##";F7
790      STOP
```

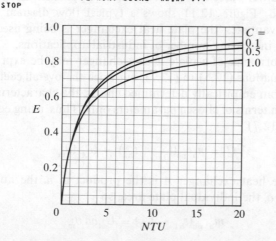

Fig. Example 12.3. Thermal Wheel characteristics.

12.5 Plate Heat Exchangers

The plate type of heat exchanger is basically of the in-line type, but the construction is very different from the conventional shell and tube concept. A plate heat exchanger consists of a frame in which a number of heat-transfer plates are supported and clamped between a header and a follower. Each plate has four ports and the edges of the plates and ports are sealed by gaskets so that hot and cold fluids flow in alternate passages formed between the plates. This means the fluids flow in very thin streams having a high heat-transfer

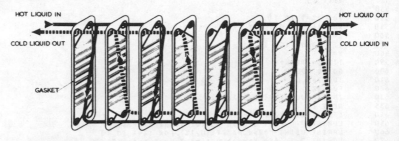

Fig. 12.11. A typical flow diagram of a plate heat exchanger showing a two-pass arrangement (diagram by courtesy of the A.P.V. Company Ltd).

area, and corrugations on the plates promote turbulence and very high heat-transfer rates. Since the plates are usually arranged for general counter-current flow, very close approach temperatures are obtained. Figure 12.11 shows a typical flow diagram. Because of these advantages, the plate heat exchanger is being used extensively in an increasing number of industrial applications.

The performance of a plate heat exchanger may be expressed in terms of equations (12.5) to (12.7), but since the overall coefficient is obtained from empirically determined charts, the characteristics are expressed in terms of chosen parameters only. Thus, using equations (12.5) and (12.7)

$$Q = m_h c_{ph}(t_{h1} - t_{h2}) = U_A A \theta_m$$

For a plate heat exchanger A is the product of n, the number of plates, and a, the individual plate area, so

$$m_h c_{ph}(t_{h1} - t_{h2}) = U_A na\, \theta_m$$

Fig. 12.12. *A Paraflow type R145 plate heat exchanger, capable of accepting up to 955 m³ per hour at 10·7 bar, and up to 130° C; plate size is 2122 × 849 mm (photograph courtesy of the A.P.V. Company Ltd).*

$$\therefore \quad n = \frac{m_h c_{ph}}{PN} \cdot \frac{(t_{h1} - t_{h2})}{\theta_m}$$

where PN is the plate number, $U_A a$. For $m_h c_{ph}$ being the minimum capacity rate, or for equal rates as defined previously, it is seen from equations (12.19), (12.5), and (12.7) that $(t_{h1} - t_{h2})/\theta_m = NTU$, the number of transfer units, and hence

$$n = \frac{m_h c_{ph}}{PN} \times NTU \qquad (12.38)$$

The performance of a particular plate design can be expressed graphically in terms of the plate number, the NTU value, and the pressure drop plotted against the plate rate, or the mass flow rate across a plate, see Fig. 12.13. Separate curves would exist for different capacity ratios, and from such information for various plate designs, the required unit for a particular duty can be selected. Certain correction factors have to be introduced, on account of concurrency and other effects which depend upon the particular plate arrangement, and on account of uneven distribution along the plate pack due to pressure losses along the ports. For exactness liquid properties have also to be considered, and separate relationships would apply to laminar and transitional flow.

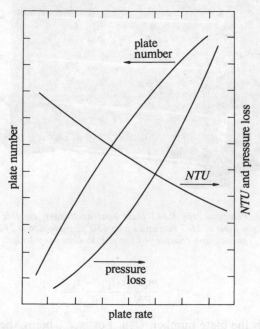

Fig. 12.13. Plate number, pressure loss and NTU characteristics of a plate heat exchanger for turbulent flow.

12.6 Batch Heat Exchangers

In certain industries where a fluid to be heated is of limited amount, or alternatively is of a very viscous nature, as for example in food processing, the batch heat exchanger is used. The entire mass of fluid to be heated is contained in a stirred vessel and is in thermal contact with a constant temperature source of heat, such as a coiled pipe through which steam is passing. The mass of fluid is essentially of uniform temperature at any given time, and the theory of Section 5.1 applies. Figure 12.14 shows the temperature time curve for the heat exchange process, and from equation (5.1), the temperature–time curve is given by:

$$\theta_2/\theta_1 = \exp(-t/\mathbf{T}) \qquad (12.39)$$

where t is the heating time between temperature differences of θ_1 and θ_2 and $\mathbf{T}$ is the time constant mC_p/hA, where m is the mass of heated and stirred fluid, C_p is the specific heat, h is the convection coefficient between the pipe and fluid, and A is the area of the heating surface (e.g., $A = \pi dL$ for a pipe) in contact with the heated fluid.

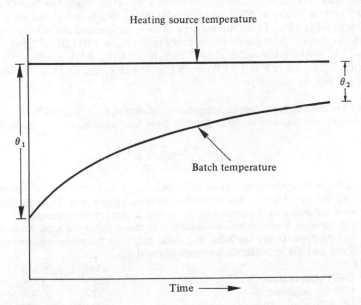

Fig. 12.14. Temperature–time curves for batch heat exchanger.

PROBLEMS

1. A tubular heater of the counter flow type is used to heat 1·26 kg/s of fuel oil of specific heat 3·14 kJ/(kg K) from 10° to 26·7°C. Heat is supplied by means of 1·51 kg/s of water which enters the heater at 82°C.

(a) Derive an equation relating the temperatures of oil and water at any section of the heater.

(b) Determine the necessary surface if the rate of heat transfer is 1·135 kW/(m² K). (Ans.: 1·013 m²) (*University College, London*).

2. In a test on a steam condenser the rate of flow of cooling water was varied whilst the condensation temperature was maintained constant. The following results were obtained:

Overall heat transfer coefficient K, kW/(m² K)	2·7	2·98	3·39	3·59
Water velocity V, m/s	0·986	1·27	1·83	2·16

Assuming the surface coefficient on the water side to be proportional to $V^{0·8}$, determine from an appropriate graph, the mean value of the steam side surface coefficient. The thickness of the metal wall is 0·122 cm and thermal conductivity of tube material 0·111 kW/(m K). (Ans.: 6·04 kW/(m² K.) (*University of Manchester*).

3. A counter flow heat exchanger consists of a bundle of 20 mm diameter tubes contained in a shell. Oil flowing in the tubes is cooled by water flowing in the shell. The flow area within the tubes is $4·4 \times 10^{-3}$ m². The flow of oil is 2·5 kg/s; it enters at 65°C and leaves at 48°C. Water enters the shell at 2·0 kg/s and at 15°C. Calculate the area of tube surface and the effectiveness of the exchanger. For the oil in the tubes take $Nu_d = 0·023 \, (Re_d)^{0·8}(Pr)^{0·33}$, $c_p = 2·15$ kJ/(kg K), $\mu = 2·2 \times 10^{-5}$ Pa s, $\rho = 880$ kg/m³, $k = 190 \times 10^{-6}$ kW/(m K); for water $\overline{h} = 1·2$ kW/(m² K), $c_p = 4·19$ kJ/(kg K). (Ans.: 2·61 m², 34%.) (*The City University*).

4. (i) Define the term 'mean temperature difference' as applied to a heat exchanger and show that, for a counter flow heat exchanger, it is given by

$$\Delta t_m = \frac{\Delta t_2 - \Delta t_1}{\ln (\Delta t_2 / \Delta t_1)}$$

where Δt_m is the mean temperature difference, Δt_1 is the temperature difference between the two fluids at one end of the heat exchanger, and Δt_2 is the temperature difference at the other end. State any necessary assumptions.

(ii) A tubular, counter flow oil cooler is to use a supply of cold water as the cooling fluid. Using the following data, calculate the mean temperature difference and the required surface area of the tubes.

Data:	Oil	Water
Entry temperature, °C	121	15·6
Exit temperature, °C	82·3	—
Mass flow rate, kg/s	0·189	0·378
Specific heat, kJ/(kg K)	2·094	4·187

Mean overall coefficient of heat transfer, referred to outside surface of tubes, 0·454 kW/(m² K). (Ans: 80·0 K, 0·422 m².) (*Imperial College, London*).

5. Two counter flow heat exchanger schemes are shown in the diagrams. In each scheme it is required to cool a fluid from 140° to 90°C using a counter flow rate of water of 1·2 kg/s entering at 30° and leaving at 80°C. In scheme (b) each unit takes half the flow of the fluid. The overall heat-transfer coefficient is 0·9 kW/(m² K) in both cases. Calculate the total area of heat exchange surface in each case, assuming a capacity ratio of 1. (Ans.: (a) 4·65 m², (b) 4·83 m².) (*The City University*).

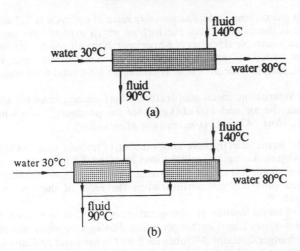

6. An industrial fluid is cooled by oil in a parallel flow heat exchanger, from 280° to 160°C while the oil enters at 64° and leaves at 124°C. Find the minimum temperature to which the oil could be cooled in parallel flow and also in counter flow for the same entry temperatures. Find the ratio of heat exchange area in parallel flow to that in counter flow, for an outlet fluid temperature of 160°C. (Ans.: 136°C, 64°C, 1·23 to 1.)

7. An oil cooler consists of a straight tube, of inside diameter 1·27 cm, wall thickness 0·127 cm enclosed within a pipe and concentric with it. The external surface of the pipe is well lagged.

Oil flow through the tube at the rate of 0·063 kg/s and cooling water flows in the annulus between the tube and the pipe at the rate of 0·0756 kg/s and in the direction opposite to that of the oil. The oil enters the tube at 177°C and is cooled to 65·5°C. The cooling water enters at 10°C.

Estimate the length of tube required, given that the heat transfer coefficient from oil to tube surface is 1·7 kW/(m² K), and that from the surface to water is 3·97 kW/(m² K). Neglect the temperature drop across the tube wall. The specific heat of the oil is 1·675 kJ/(kg K). (Ans.: 2·67 m.) (*University of London*).

8. A tank contains 272 kg of oil which is stirred so that its temperature is uniform. The oil is heated by an immersed coil of pipe 2·54 cm diameter in which steam condenses at 149°C. The oil, of specific heat 1·675 kJ/(kg K) is to be heated from 32·2° to 121°C in 1 hour. Calculate the length of pipe in the coil if the surface coefficient is 0·653 kW/(m² K). (Ans.: 3·47 m.)

9. Explain briefly what is meant by the term 'surface or film coefficient' in heat transfer considerations.

A counter-flow heat exchanger having an overall heat transfer coefficient of 0·114 kW/(m² K) is used to heat to 329°C the air entering the combustion chamber of a gas turbine cycle. The pressure ratio of the cycle is 5:1 and the heating fluid is the exhaust from the turbine which expands the gas from 650°C with an isentropic efficiency of 82 per cent. If the air conditions initially are 1·013 bar and 21°C and the isentropic efficiency of the compressor is 80 per cent, calculate the area of heat exchanger for a total fluid mass flow of 22·7 kg/s.

Assume a logarithmic mean temperature difference and constant specific heat of 1·0 for the air and 1·09 kJ/(kg K) for the products. $\gamma = 1·4$ for air and products. (Ans.: 424 m^2.) (*University of Manchester*).

10. Define the terms 'effectiveness' and 'number of transfer units' as applied to heat exchangers stating any assumptions involved. Obtain a relationship between effectiveness and number of transfer units for a counter-current heat exchanger and plot this relationship when the ratio of the stream heat capacities is 0·5.

20·15 kg/s of an oil fraction at a temperature of 121°C is to be cooled in a simple counter-current heat exchanger using 5·04 kg/s of water initially at 10°C. The exchanger contains 200 tubes each 4·87 m long and 1·97 cm outside diameter; the resulting heat transfer coefficient referred to the outside tube area is 0·34 kW/(m^2 K). If the specific heat of the oil is 2·094 kJ/(kg K) calculate the exit temperature of the oil. (Ans.: 90·8°C.) (*University of Leeds*).

11. Write a program in BASIC to consider heat exchange in steady state for a single pass unmixed flow cross-flow heat exchanger.

Determine the Effectiveness for Capacity Ratios of value 0·25, 0·5, 0·75 and 1·0 for *NTU* values up to 5, and compare with analytical results given by Kays and London.[3]

12. Write a program in BASIC to consider transient response of a single pass unmixed flow cross-flow heat exchanger. Consider two versions of the program, based on equations (12.27) and (12.28). Obtain steady state effectiveness results from the previous question for a Capacity Ratio of 1·0, for *NTU* values of 1 and 5. Then determine and plot the percentage of steady state effectiveness achieved at given real times up to 1200 seconds, for fluid to wall capacity ratios of 5·0 and 2·5 for *NTU* values of 1 and 5. Study the effect of using the two different equations, (12.27) and (12.28). The result of this analysis is given in Ref. 4.

REFERENCES

1. Smith, D. M. *Engineering*, Vol. 138, 479, 606 (1934).
2. Bowman, R. A., Mueller, A. C., and Nagle, W. M. *Trans. ASME*, Vol. 62, 283 (1940).
3. Kays, W. M., and London, A. L. *Compact Heat Exchangers*, McGraw-Hill Book Company, Inc., New York (1964).

4. Simonson, J. R. 'Transient and steady state analysis of cross flow heat exchangers by programs in Fortran', *Trans. Inst. Chem. Engrs.*, Vol. 55, 53–58 (1977).

13

The laws of black- and grey-body radiation

The processes of heat transfer considered so far have been intimately related to the nature of the material medium, the presence of solid–fluid interfaces, and the presence of fluid motion. Energy transfer has been observed to take place only in the direction of a negative temperature gradient, and at a rate which depends directly on the magnitude of that gradient.

It is now necessary to consider the third mode of heat transfer which is characteristically different from conduction and convection. Radiation occurs most freely in a vacuum, it is freely transmitted in air (though partially absorbed by other gases) and, in general, is partially reflected and partially absorbed by solids. Transmission of radiation, which can occur in solids as well as fluids, is an interesting phenomenon because it can occur through a cold non-absorbing medium between two other hotter bodies. Thus the surface of the earth receives energy direct by radiation from the sun, even though the atmosphere at high altitude is extremely cold. Similarly, the glass of a green house is colder than the contents and radiant energy does not stop there, it is transmitted to the warmer absorbing surfaces inside. Radiation is also significantly different from conduction and convection in that the temperature level is a controlling factor. In furnaces and combustion chambers, radiation is the predominating mechanism of heat transfer.

As already mentioned in Chapter 1, radiant energy is but part of the entire spectrum of electromagnetic radiation. All radiation travels at the speed of light and, consequently, longer wave-lengths correspond to lower frequencies, and shorter wave-lengths to higher frequencies. The entire spectrum of electromagnetic radiation extends from about 10^{-4} angstrom units (10^{-14} metres), the wavelength region of cosmic rays, up to about 20,000 metres, in the region of Hertzian or electric waves. The wave-length region generally associated with thermal radiation is 10^3–10^6 angstrom units, which includes some ultra-violet, all the visible, and some infra-red radiation. Figure 13.2 shows part of the spectrum of electromagnetic radiation.

Since radiation energy exchange depends on the rates at which energy is emitted by one body and absorbed by another, it is necessary to establish definitions relating to these characteristics of surfaces. Further, not all of the energy emitted by one body may necessarily fall on the surface of another due to their geometric arrangement, and this too must be investigated. This then forms the general approach by which engineers may consider radiant energy exchange.

13.1 Absorption and Reflection of Radiant Energy

Three possibilities may follow the incidence of radiation on the surface of a body. Some may be transmitted through the body leaving it unaltered. Some may be absorbed on the surface, resulting in an increase in temperature of the body at the surface. The remainder will have been reflected. This can take place in two ways, either as *specular* reflection where the angle of reflection is equal to the angle of incidence, or as *diffuse* reflection where the reflected energy leaves in all directions from the surface. Thus polished surfaces tend to be specular and rough surfaces diffuse.

The percentage of incident energy absorbed by a surface is defined as α, the absorptivity; the percentage reflected is ρ, the reflectivity, and the percentage transmitted is τ, the transmissivity. Thus it must follow that

$$\alpha + \rho + \tau = 1 \qquad (13.1)$$

Energy absorbed on the surface is, in fact, absorbed in a finite thickness of material, and if the body is very thin less absorption and more transmission may take place. It will be assumed that 'thick' bodies only will be considered, for which $\tau = 0$. Hence

$$\alpha + \rho = 1 \qquad (13.2)$$

In engineering applications of radiation, there will generally be a gas separating solid bodies, and often this gas is air which may be assumed to have no absorptivity or reflectivity, so $\tau = 1$. Combustion gases containing carbon dioxide and water vapour behave very differently, however, and an elementary treatment of non-luminous gas radiation appears later in this chapter.

13.2 Emission, Radiosity, and Irradiation

To be consistent with previous nomenclature, Q is the energy emitted by a surface in heat units per unit time. This energy emission results from the surface temperature and the nature of the surface. However, Q may not be the total energy leaving that surface, there may also be some reflected incident energy. Thus J is defined as the Radiosity, which is the total radiant energy leaving the surface, in unit time. Similarly, G is defined as the Irradiation which is total incident energy on a surface, some of which may be emission and some reflection from elsewhere.

If G is the incident energy, ρG will be reflected. Thus

$$J = Q + \rho G \tag{13.3}$$

13.3 Black and Non-black Bodies

All materials have values of α and ρ between 0 and 1. However, it is useful and important to imagine a material for which $\alpha = 1$ and $\rho = 0$. A body composed of this material is known as a black body; it absorbs all incident energy upon it and reflects none. For real materials the highest values of α are around 0·97. Artificial surfaces may be arranged in practice which are virtually black. Consider Fig. 13.1. The hollow enclosure has an inside surface of high absorptivity. Incident energy passes through the small opening and is

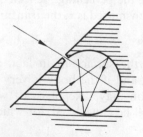

Fig. 13.1.   Artificial black-body surface.

absorbed on the inside surface. However, some is reflected, but most of this is absorbed on a second incidence. Again, a small fraction is reflected. After a number of such reflections the amount unabsorbed is exceedingly small and very little of the original incident energy is reflected back out of the opening. The area of the opening may thus be regarded as black.

The work of Stefan and Boltzmann led to the law named after them which gives the emission of radiant energy from a black body. Thus

$$Q_b = A\sigma T^4 \quad \text{or} \quad q_b = \sigma T^4 \tag{13.4}$$

is the Stefan–Boltzmann law for black-body radiation. T is the absolute temperature and σ is the Stefan–Boltzmann constant and has the value $56{\cdot}7 \times 10^{-12}\,\text{kW/(m}^2\,\text{K}^4)$. A derivation of this law is given by Jakob.[1]

Black-body radiation consists of emission over the entire range of wave-length. Most of the energy is concentrated in the wave-length range already mentioned. The point to note is that the energy is not distributed uniformly over this range. Thus $q_{b\lambda}$ may be defined as the monochromatic emittance, the energy emitted per unit area at the wave-length λ, for a black body. It must follow that

$$q_b = \int_0^\infty q_{b\lambda}\,d\lambda = \sigma T^4 \tag{13.5}$$

The variation of $q_{b\lambda}$ with wavelength was established by Planck[2] in his quantum theory of electromagnetic radiation, thus

$$q_{b\lambda} = \frac{C_1 \lambda^{-5}}{\exp\left(C_2/\lambda T\right) - 1}$$

where λ = wavelength, µm, T = absolute temperature, $C_1 = 3{\cdot}743 \times 10^5\,\text{kW}\mu^4/\text{m}^2$, $C_2 = 1{\cdot}439 \times 10^4\,\mu\text{K}$. The form of the variation of $q_{b\lambda}$ is shown in Fig. 13.2, and it is seen that there is a peak value of $q_{b\lambda}$ which occurs at a wave length which is related to the absolute temperature by Wien's displacement law:

$$\lambda_{max} T = 2897{\cdot}6\,\mu\,\text{K}$$

Real materials that are not black will have monochromatic emittances that are different from $q_{b\lambda}$, and hence it is useful to define a monochromatic emissivity ε_λ by the equation

$$q_\lambda = \varepsilon_\lambda q_{b\lambda}$$

or

$$\varepsilon_\lambda = \frac{q_\lambda}{q_{b\lambda}} \tag{13.6}$$

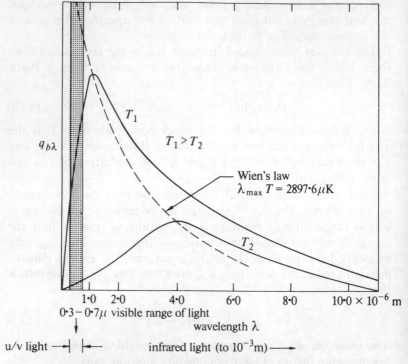

Fig. 13.2. Variation of black-body emissive power $q_{b\lambda}$ with wave-length and temperature

The black and non-black emittances which give ε_λ are measured at the same temperature. In general, ε_λ is a function of wave-length, temperature and direction. Real surfaces often exhibit directional variation in emissive power, thus non-electrically conducting materials emit more in the normal direction whereas for conducting materials often the reverse is true. For practical calculations, quoted emissivities are total hemispherical values. Most real materials exhibit some variation in ε_λ with wave-length. These are known as selective emitters. However, there is a second type of ideal surface, known as a grey surface, where the emissivity is constant with wave-length. Some real materials approximate closely to this ideal, but the concept reduces calculations to the extent that it is worthwhile to accept the error introduced in exchange for the simplifica-

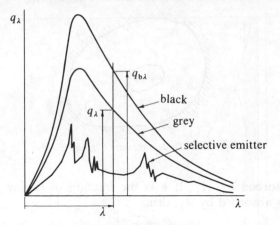

Fig. 13.3. Comparison of the emission of black, grey, and selective emitting surfaces; $\epsilon\lambda = q\lambda/q_{b\lambda}$.

tion afforded. Both grey-body and selective emission are shown in Fig. 13.3. It must follow that for a grey body

$$q = \varepsilon q_b = \varepsilon\sigma T^4 \tag{13.7}$$

The value of ε used for a grey body is generally a function of the temperature of the surface, but again a simplifying assumption enables a suitable constant value to be used, irrespective of temperature, provided the range is not too large. Values of ε for real materials, and the temperatures at which they are valid, are given in Table A.7 (see page 254).

It is now apparent that materials exist for which $\alpha < 1$ and also for which the emission is not equal to the black-body emission. By means of Kirchhoff's law the relationship between α and ε may be established.

13.4 Kirchhoff's Law[3]

Consider a small black body of area A_1 completely enclosed by a larger body with an internal black surface area A_2, as in Fig. 13.4. Both surfaces are at the same temperature. The small body will emit at the rate $A_1\sigma T^4$ and must also absorb energy at the same rate otherwise the temperature of the body will change. The concave surface A_2 will emit $A_2\sigma T^4$, but only $A_1\sigma T^4$ of this is incident upon,

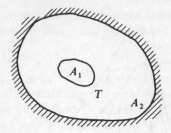

Fig. 13.4. *To demonstrate Kirchhoff's law.*

and absorbed by, A_1. If F is the fraction of energy leaving A_2 which is absorbed by A_1, then

$$F = \frac{A_1\sigma T^4}{A_2\sigma T^4} = \frac{A_1}{A_2} \qquad (13.8)$$

The remainder of the energy emitted by A_2 will be re-absorbed by A_2 as it will miss A_1.

Now consider what happens when the black body of area A_1 is replaced by a grey body of identical dimensions, with an absorptivity of α and an emissivity of ε, the temperature throughout remaining at T. Since there is again thermal equilibrium the energy actually absorbed on A_1 must equal the energy emitted by A_1. The energy emitted by A_2 is $A_2\sigma T^4$ and this is also the radiosity of A_2 since nothing is reflected by A_2. Of this, only $FA_2\sigma T^4$ will fall on A_1 and only $\alpha FA_2\sigma T^4$ will be absorbed. A_1 will itself emit $\varepsilon A_1\sigma T^4$ and this must equal the energy absorbed.

$$\therefore \qquad \varepsilon A_1\sigma T^4 = \alpha FA_2\sigma T^4$$

But

$$FA_2 = A_1 \qquad \text{from (13.8)}$$

Therefore

$$\varepsilon = \alpha \qquad (13.9)$$

Thus, Kirchhoff's law, as stated by equation (13.9), says that the absorptivity is equal to the emissivity at any given temperature. It follows that for a black body for which $\alpha = 1$, that $\varepsilon = 1$ and, consequently, $\varepsilon < 1$ for a grey body. Since it is possible to use a suitable value of ε for grey bodies over a temperature range, the value of α over that range is the same. This does not hold for real materials that are true selective emitters when the temperature difference is very large, because the bulk of the energy absorbed by either body is in a very different wave-length region than the energy emitted by that body.

13.5 Intensity of Radiation

The radiation from a unit area of black body is $q_b = \sigma T^4$. For diffuse radiation from a small flat area of black surface dA, the entire emittance Q_b must pass through a hemispherical surface surrounding the emitting area. It is necessary to consider the distribution of radiant energy per unit area over the spherical surface, before calculations can be made of radiation exchanges.

The intensity of black-body radiation, I, is the radiation emitted per unit time and unit solid angle subtended at the source, and per unit area of emitting surface normal to the mean direction in space, and may be expressed as

$$I = \frac{dQ_b}{(dA_2/r^2)\,dA_1 \cos \phi} \tag{13.10}$$

This is shown in Fig. 13.5. dA_2/r^2 is the solid angle subtended by dA_2. The radiant energy per unit area at the hemispherical surface is the *radiant flux* dQ_b/dA_2. The surface of dA_1 has been specified as diffuse, thus Lambert's law[4] states that I is constant in the hemispherical space above dA_1. From the above definition of I it thus

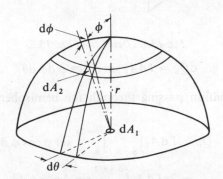

Fig. 13.5. To evaluate intensity of radiation.

follows that dQ_b/dA_2 will have a maximum value at any given r when $\phi = 0$, i.e., when dA_2 is on the normal to dA_1. Further, dQ_b/dA_2 is zero when $\phi = 90°$ and, in addition, dQ_b/dA_2 will vary inversely as r^2. In general,

$$\left(\frac{dQ_b}{dA_2}\right)_\phi = \left(\frac{dQ_b}{dA_2}\right)_n \cos \phi$$

where the suffix n implies on the normal to dA_1.

For Lambert's law to be true, I for a black surface must depend on the absolute temperature only. From equation (13.10),

$$dQ_b = I\left(\frac{dA_2}{r^2}\right) dA_1 \cos \phi \qquad (13.11)$$

and from Fig. 13.6 it is seen that $dA_2 = r d\phi\,(r \sin \phi\, d\theta) = r^2 \sin \phi\, d\phi\, d\theta$. Hence

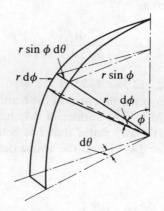

Fig. 13.6. Detail from Fig. 13.5.

$$dQ_b = I\, dA_1 \sin \phi \cos \phi\, d\phi\, d\theta$$

The total radiation passing through the hemispherical surface is then

$$Q_b = I\, dA_1 \int_{\phi=0}^{\phi=\pi/2} \int_{\theta=0}^{\theta=2\pi} \sin \phi \cos \phi\, d\phi\, d\theta$$

$$= 2\pi I\, dA_1 \int_{\phi=0}^{\phi=\pi/2} \sin \phi \cos \phi\, d\phi$$

$$= \pi I \, dA_1$$

$$\therefore \quad I = \frac{q_b}{\pi} = \frac{\sigma T^4}{\pi} \tag{13.12}$$

13.6 Radiation Exchange between Black Surfaces

It is now possible to consider the radiation exchange between two arbitrarily disposed black surfaces of area A_1 and A_2, and at temperatures T_1 and T_2. Small elements of each surface dA_1 and dA_2 are considered as shown in Fig. 13.7. They are distance r

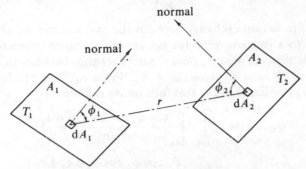

Fig. 13.7. Arbitrarily disposed black surfaces exchanging radiation.

apart, and the line joining their centres makes angles ϕ_1 and ϕ_2 to their normals. Each element of area subtends a solid angle at the centre of the other; these are $d\omega_1$ subtended at dA_1 by dA_2, and $d\omega_2$ subtended at dA_2 by dA_1. The solid angles are given by:

$$d\omega_1 = \frac{dA_2 \cos \phi_2}{r^2}, \quad \text{and} \quad d\omega_2 = \frac{dA_1 \cos \phi_1}{r^2}$$

From equation (13.11) the radiant energy emitted by dA_1 that impinges on dA_2 is given by:

$$dQ_{b(1-2)} = I_1 \, dA_1 \cos \phi_1 \left(\frac{dA_2 \cos \phi_2}{r^2} \right) \tag{13.13}$$

Since both surfaces are black this energy is absorbed by dA_2. A similar quantity of energy is also radiated by dA_2 and absorbed by dA_1 expressed as

$$dQ_{b(2-1)} = I_2 \, dA_2 \cos \phi_2 \left(\frac{dA_1 \cos \phi_1}{r^2} \right) \tag{13.14}$$

The net exchange is

$$dQ_{b(1-2)} - dQ_{b(2-1)} = dQ_{b(12)}$$

and

$$dQ_{b(12)} = \frac{dA_1\, dA_2 \cos \phi_1 \cos \phi_2}{r^2}(I_1 - I_2)$$

Equation (13.12) is now used to give the final result

$$dQ_{b(12)} = \frac{\sigma dA_1\, dA_2 \cos \phi_1 \cos \phi_2}{\pi r^2}(T_1^4 - T_2^4) \qquad (13.15)$$

The total radiation exchange between the two surfaces A_1 and A_2 amounts to a summation of the net energy exchange between dA_1 and all elements of area A_2, and the net exchange between all other elements of A_1 and all elements of A_2. From equation (13.13), the total energy radiated by A_1 that falls on A_2 is given by

$$Q_{b(1-2)} = I_1 \int_{A_1} \int_{A_2} \frac{\cos \phi_1 \cos \phi_2\, dA_1\, dA_2}{r^2}$$

$$= \sigma T_1^4 \int_{A_1} \int_{A_2} \frac{\cos \phi_1 \cos \phi_2\, dA_1\, dA_2}{\pi r^2}$$

But the total energy radiated by A_1 is

$$Q_{b(1)} = A_1 \sigma T_1^4$$

Hence the fraction of energy radiated by A_1 that falls on A_2 is

$$\frac{Q_{b(1-2)}}{Q_{b(1)}} = \frac{1}{A_1} \int_{A_1} \int_{A_2} \frac{\cos \phi_1 \cos \phi_2\, dA_1\, dA_2}{\pi r^2}$$

$$= F_{1-2} \qquad (13.16)$$

F_{1-2} is known as the geometric configuration factor of A_1 with respect to A_2. Thus the energy radiated by A_1 that falls on A_2 may be expressed as

$$Q_{b(1-2)} = F_{1-2}A_1 \sigma T_1^4 \qquad (13.17)$$

Similarly, from equation (13.14) the total energy radiated by A_2 that falls on A_1 is given by

$$Q_{b(2-1)} = \sigma T_2^4 \int_{A_1} \int_{A_2} \frac{\cos \phi_1 \cos \phi_2\, dA_1\, dA_2}{\pi r^2}$$

and the total energy radiated by A_2 is $A_2\sigma T_2^4$, so that

$$\frac{Q_{b(2-1)}}{Q_{b(2)}} = \frac{1}{A_2} \int_{A_1} \int_{A_2} \frac{\cos\phi_1 \cos\phi_2\, dA_1\, dA_2}{\pi r^2}$$

$$= F_{2-1} \qquad (13.18)$$

and

$$Q_{b(2-1)} = F_{2-1} A_2 \sigma T_2^4 \qquad (13.19)$$

From equations (13.16) and (13.18) it is seen that F_{1-2} and F_{2-1} are simply related:

$$A_1 F_{1-2} = A_2 F_{2-1} \qquad (13.20)$$

The net radiation exchange from equations (13.17) and (13.19) can be expressed in terms of either configuration factor, thus

$$Q_{b(12)} = F_{1-2} A_1 \sigma (T_1^4 - T_2^4)$$

$$= F_{2-1} A_2 \sigma (T_1^4 - T_2^4) \qquad (13.21)$$

It is necessary to know or to be able to calculate configuration factors before black-body radiation exchanges can be determined. Only a few results will be considered here, and the reader is referred elsewhere for further information on this subject.[1,5,6]

13.6.1 Examples of the Black-Body Geometric Configuration Factor

(i) *Cases where* $F_{1-2} = 1$. The simplest case is when surface A_1 is entirely convex and is completely enclosed by A_2. Then F_{1-2} must be 1, since all the energy radiated by A_1 must fall on A_2. It follows also that F_{2-1} is A_1/A_2. In this case, the net black-body radiation exchange is

$$Q_{b(12)} = A_1 \sigma (T_1^4 - T_2^4) \qquad (13.22)$$

Another simple example is when surfaces A_1 and A_2 are parallel and large, and radiation occurs across the gap between them, so that in this case $A_1 = A_2$ and all radiation emitted by one falls on the other if edge effects are neglected. Hence,

$$F_{1-2} = F_{2-1} = 1$$

Concentric surfaces may be included if the gap between them is small so that little error is introduced by the small difference between the area of A_1 and A_2. The net radiation exchange is again given by equation (13.22).

(ii) *Small arbitrarily disposed areas.* In some circumstances it is possible to use equation (13.15) as it stands, if the areas dA_1 and dA_2 are small. Thus the energy received by a small disc placed in front of a small window in a furnace could be approximately calculated this way.

(iii) *Thermocouple in a circular duct.* A simple practical example of the geometric configuration factor is found in consideration of a thermocouple in a circular duct. It may be assumed that the thermocouple joint is represented by a small sphere and, further, that it is situated at the centre of a duct of length $2L$ and radius R. It is illustrated in Fig. 13.8. The line joining elements of area always

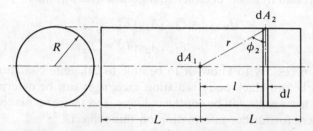

Fig. 13.8. The thermocouple configuration factor.

strikes the thermocouple joint normally, so $\cos \phi_1$ is always 1. The element of area of the duct wall is $2\pi R\, dl$. Since A_1 is a very small sphere of radius r_c, dA_1 is the disc area πr_c^2, and is constant.

Applying equation (13.16) gives

$$\frac{Q_{b(1-2)}}{Q_{b(1)}} = \frac{dA_1}{A_1} \int_{A_2} \frac{\cos \phi_2\, 2\pi R\, dl}{\pi r^2}$$

But $\cos \phi_2 = R/r$ and $r = (R^2 + l^2)^{\frac{1}{2}}$

$$\therefore \quad \frac{Q_{b(1-2)}}{Q_{b(1)}} = \frac{\pi r_c^2}{4\pi r_c^2} \int_{-L}^{+L} \frac{2R^2\, dl}{(R^2 + l^2)^{\frac{3}{2}}}$$

$$= \frac{1}{4}\left[\frac{2l}{(R^2 + l^2)^{\frac{1}{2}}} \right]_{-L}^{+L} = \frac{L}{(R^2 + L^2)^{\frac{1}{2}}} \quad (13.23)$$

EXAMPLE 13.1

A thermocouple situated at the centre of a circular duct 10 cm diameter by 0·25 m long has a spherical bead 2 mm diameter. It reads 185°C with gas at 200°C flowing along the duct; the wall of the duct is at 140°C. Determine a convection coefficient for heat transfer between the gas and the bead, assuming radiating surfaces are black.

Solution. Convection to the thermocouple from the gas is equal to the radiation exchange between the thermocouple and the wall. The configuration factor is $\dfrac{0·125}{(0·05^2 + 0·125^2)^{\frac{1}{2}}} = 0·93$. If h is the convection coefficient, and A the area of the bead, then

$$Q_b = 0·93 \times A \times 56·7 \times 10^{-4} \left[\left(\frac{458}{100}\right)^4 - \left(\frac{413}{100}\right)^4 \right] = hA\theta$$

where $\theta = 200 - 185 = 15$

$$\therefore \quad 52·7 \times 10^{-4} (441 - 292) = 15h$$

$$\therefore \quad h = 0·0523 \; kW/(m^2 \; K)$$

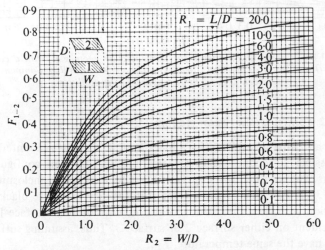

Fig. 13.9. Configuration factors for parallel opposed rectangles. (From A. J. Chapman, Heat Transfer, The MacMillan Company, New York (1974). By permission of the publisher.)

(iv) *Parallel and perpendicular rectangles.* Radiation exchanges between finite parallel rectangles and perpendicular rectangles with a common edge occur in furnaces, etc., and details of the application of equation (13.16) to these cases may be found in ref. 6. Calculated values of the configuration factor are available in graphical form, shown in Fig. 13.9 for parallel rectangles and Fig. 13.10 for perpendicular rectangles.

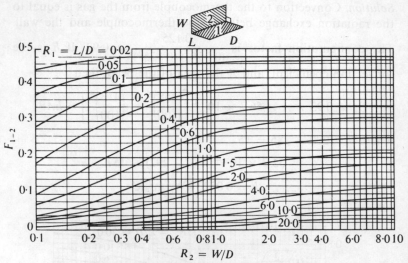

Fig. 13.10. *Configuration factor for perpendicular rectangles with a common edge. (From A. J. Chapman, Heat Transfer, The MacMillan Company, New York (1974). By permission of the publisher.)*

13.6.2 Configuration Factor Algebra

Since all radiation emitted by a black surface must be accounted for it is possible to deduce algebraic relationships between configuration factors. Suppose that three surfaces make up a complete enclosure as in Fig. 13.11. Any one of those three surfaces may exchange black body radiation with the other two, and all the emission from surface 1 must be incident on either surface 2 or surface 3. Thus, assuming surfaces 2 and 3 have the same temperature, T_2:

$$A_1 F_{1-2+3} \; \sigma \left(T_1^4 - T_2^4 \right) = A_1 F_{1-2} \; \sigma \left(T_1^4 - T_2^4 \right)$$

$$+ A_1 F_{1-3}\, \sigma\, (T_1^4 - T_2^4)$$

$$\therefore\ F_{1-2+3} = F_{1-2} + F_{1-3} \tag{13.24}$$

For any number of surfaces, the configuration factor from one to the remainder is the sum of the component factors. This result is useful since unknown factors may be deduced from known standard cases.

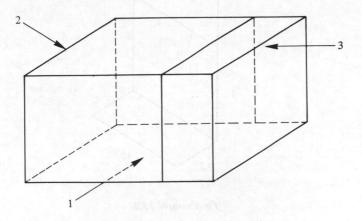

Fig. 13.11. Three surfaces making an enclosure.

EXAMPLE 13.2

It is required to find F_{1-2} for the case in Fig. 13.12, i.e., rectangles at right angles, but not sharing a common edge. The sides of the rectangles are of unit or half unit length as shown.

From Fig. 13.10, $F_{1-2+3} = 0.32$, for $W/D = 2.0$ and $L/D = 0.5$. Also $F_{1-3} = 0.29$ for $W/D = 1.0$ and $L/D = 0.5$. Hence, from equation (13.24):

$$F_{1-2} = F_{1-2+3} - F_{1-3}$$

$$\therefore\ F_{1-2} = 0.32 - 0.29 = 0.03$$

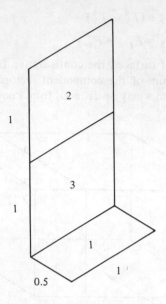

Fig. Example 13.2.

13.7 Grey-Body Radiation Exchanges

When radiating surfaces are grey, the emissivities of those surfaces must be taken into account as well as their geometric configuration. To enable the equation for a net energy exchange to be written in a similar manner to that for black-body radiation, Hottel[8] introduced a new factor $\mathscr{F}$. Thus a net exchange is expressed as

$$Q_{(12)} = A_1\mathscr{F}_{1-2}\sigma(T_1^4 - T_2^4) \tag{13.25}$$

The derivation of $\mathscr{F}$ will be considered by means of an electrical analogy of radiation.[7] In the case of a net black-body radiation exchange (13.21) is compared with Ohm's law, so that

$$Q_{b(12)} = F_{1-2}A_1\sigma(T_1^4 - T_2^4) \quad \text{is equivalent to} \quad I = \Delta V/R$$

Hence

$$Q_{b(12)} \equiv I; \qquad \sigma(T_1^4 - T_2^4) \equiv \Delta V; \qquad \text{and} \qquad \frac{1}{F_{1-2}A_1} \equiv R$$

The corresponding electric circuit is shown in Fig. 13.12.

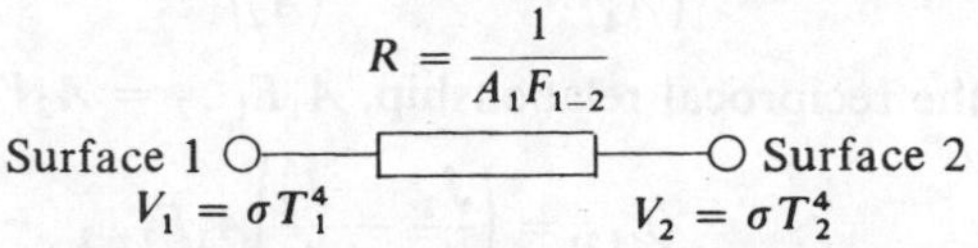

$$R = \frac{1}{A_1 F_{1-2}}$$

Surface 1 $\circ$———[   ]———$\circ$ Surface 2

$$V_1 = \sigma T_1^4 \qquad\qquad V_2 = \sigma T_2^4$$

Fig. 13.12. An equivalent electric circuit for a net black-body radiation exchange $Q_{b(12)} = A_1 F_{1-2}\, \sigma(T_1^4 - T_2^4)$.

An important initial assumption is that each radiating surface has a constant value of ρ and ε over the whole surface. From the definitions of radiosity and irradiation in section 13.2 it follows that the net rate at which energy leaves a grey surface is the difference $J - G$, and from equation (13.3)

$$J = \varepsilon Q_b + \rho G$$

$$\therefore \quad J - G = J - \frac{J - \varepsilon Q_b}{\rho}$$

and since $\rho + \varepsilon = 1$ for opaque surfaces, this reduces to

$$J - G = \frac{\varepsilon}{\rho}(Q_b - J)$$

If two surfaces only are involved, and these form an enclosure, this is also the net energy exchange between them, $Q_{(12)}$, and the equation may be compared with Ohm's law so that Q_b/A, which is σT^4, and J/A are potentials and $\rho/A\varepsilon$ is the resistance. The corresponding circuit element for either surface is shown in Fig. 13.13.

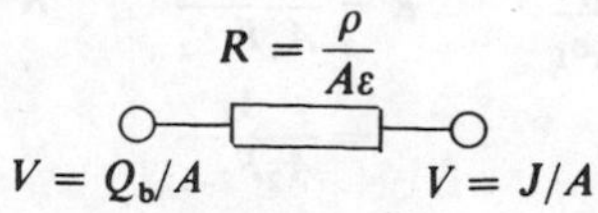

$$R = \frac{\rho}{A\varepsilon}$$

$\circ$———[   ]———$\circ$

$$V = Q_b/A \qquad\qquad V = J/A$$

Fig. 13.13.

Further, for surfaces of area A_1 and A_2 (at temperatures T_1 and T_2) which have configuration factors of F_{1-2} and F_{2-1}, the net energy exchange is also the difference between the total radiation leaving A_1 which reaches A_2, and the total radiation leaving A_2 which reaches A_1. Thus

$$Q_{(1\,2)} = \left(\frac{J_1}{A_1}\right)A_1F_{1-2} - \left(\frac{J_2}{A_2}\right)A_2F_{2-1}$$

But, from the reciprocal relationship, $A_1F_{1-2} = A_2F_{2-1}$,

$$\therefore \qquad Q_{(1\,2)} = \left(\frac{J_1}{A_1} - \frac{J_2}{A_2}\right)A_1F_{1-2}$$

This may also be represented by a circuit element, with potentials J_1/A_1 and J_2/A_2 and resistance $1/A_1F_{1-2}$, as shown in Fig. 13.14.

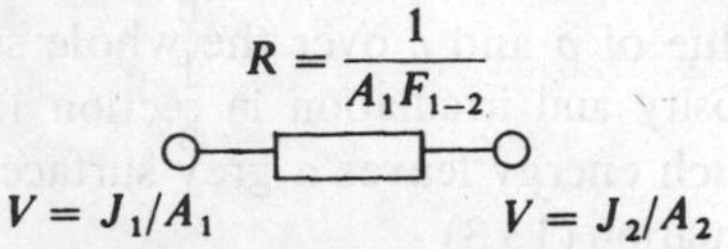

$$R = \frac{1}{A_1F_{1-2}}$$

$V = J_1/A_1$ $\qquad\qquad\qquad$ $V = J_2/A_2$

Fig. 13.14.

To simulate completely an energy exchange between the surfaces A_1 and A_2, three circuit elements may be joined in series as shown in Fig. 13.15, the whole circuit now being compared to equation (13.25). σT_1^4 and σT_2^4 are the end potentials, (equivalent to $Q_{b(1)}/A_1$ and $Q_{b(2)}/A_2$), and the total resistance is

$$\frac{\rho_1}{A_1\varepsilon_1} + \frac{1}{A_1F_{1-2}} + \frac{\rho_2}{A_2\varepsilon_2}$$

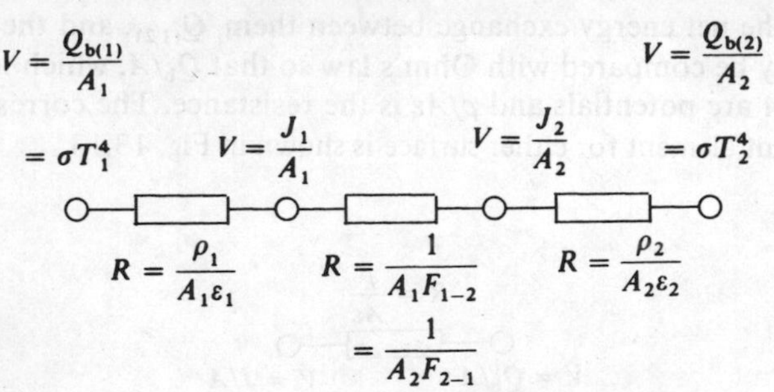

$V = \dfrac{Q_{b(1)}}{A_1}$ $\qquad\qquad\qquad\qquad\qquad$ $V = \dfrac{Q_{b(2)}}{A_2}$

$= \sigma T_1^4$ $\qquad$ $V = \dfrac{J_1}{A_1}$ $\qquad$ $V = \dfrac{J_2}{A_2}$ $\qquad$ $= \sigma T_2^4$

$R = \dfrac{\rho_1}{A_1\varepsilon_1}$ $\qquad$ $R = \dfrac{1}{A_1F_{1-2}}$ $\qquad$ $R = \dfrac{\rho_2}{A_2\varepsilon_2}$

$$= \frac{1}{A_2F_{2-1}}$$

Fig. 13.15. *Complete circuit for radiation exchange between two grey surfaces forming an enclosure.*

From equation (13.25) the resistance is also given by $1/A_1\mathscr{F}_{1-2}$, hence

$$\frac{1}{A_1\mathscr{F}_{1-2}} = \frac{\rho_1}{A_1\varepsilon_1} + \frac{1}{A_1F_{1-2}} + \frac{\rho_2}{A_2\varepsilon_2}$$

Multiplying both sides by A_1 and substituting $1 - \varepsilon$ for ρ gives

$$\frac{1}{\mathscr{F}_{1-2}} = \left(\frac{1 - \varepsilon_1}{\varepsilon_1}\right) + \frac{1}{F_{1-2}} + \frac{A_1}{A_2}\left(\frac{1 - \varepsilon_2}{\varepsilon_2}\right)$$

$$\therefore \quad \mathscr{F}_{1-2} = \frac{1}{\left(\dfrac{1}{\varepsilon_1} - 1\right) + \dfrac{1}{F_{1-2}} + \dfrac{A_1}{A_2}\left(\dfrac{1}{\varepsilon_2} - 1\right)} \quad (13.26)$$

This result may be used for any two surfaces of area A_1 and A_2, provided they form an enclosure, for which the configuration factor is F_{1-2}. Equation (13.26) simplifies for the special case of infinite parallel or concentric grey planes for which $F_{1-2} = 1$ and $A_1 = A_2$. Then,

$$\mathscr{F}_{1-2} = \frac{1}{\dfrac{1}{\varepsilon_1} + \dfrac{1}{\varepsilon_2} - 1} \quad (13.27)$$

This result can also be readily achieved without reference to the equivalent electric circuit.

A further simple result which is useful is that if A_1 is completely enclosed by A_2, so that $F_{1-2} = 1$, and A_2 is large compared with A_1 so that

$$\frac{A_1}{A_2}\left(\frac{1}{\varepsilon_2} - 1\right) \approx 0$$

then equation (13.26) reduces to

$$\mathscr{F}_{1-2} = \varepsilon_1 \quad (13.28)$$

EXAMPLE 13.3

A small oven measures 0·4 m by 0·5 m by 0·3 m high. The floor of the oven receives radiation from all the walls and roof which are at 300°C and have an emissivity of 0·8. The floor is maintained at 150°C and has an emissivity of 0·6. Calculate the radiation exchange.

Solution. A_1 is the total area of walls and roof, which is 0·74 m².

A_2 is the area of the floor, 0·2 m². Since A_1 encloses A_2, $F_{2-1} = 1$ and $F_{1-2} = A_2/A_1 = 0\cdot27$.

The grey-body factor $\mathscr{F}_{1-2}$ is $\dfrac{1}{\left(\dfrac{1}{0\cdot8} - 1\right) + \dfrac{1}{0\cdot27} + 3\cdot7\left(\dfrac{1}{0\cdot6} - 1\right)}$

$$= 0\cdot156$$

The radiation exchange is

$$0\cdot156 \times 0\cdot74 \times 56\cdot7 \times 10^{-4}\left[\left(\frac{573}{100}\right)^4 - \left(\frac{423}{100}\right)^4\right] = 0\cdot495\,\text{kW}$$

13.8 Non-luminous Gas Radiation

The more simple gas molecules such as the monatomic gases, and diatomic gases of symmetric structure such as oxygen and nitrogen, are effectively transparent to thermal radiation. However other gases and vapours are good absorbers and emitters, and gas radiation plays an important part in many practical heat transfer calculations.

The absorptive properties of a gas layer or volume depend on the wave-lenth of the incident radiation, and on a function of the number of gas molecules in the path length, expressed as a function of path length and partial pressure. Thus

$$dI_\lambda = -I_\lambda a_\lambda p\,dx$$

expresses the reduction of intensity of the monochromatic incident beam in passing through a distance dx at partial pressure p, where a_λ is the absorption coefficient. If I_λ has the value $I_{\lambda0}$ at $x = 0$, this will integrate to give

$$I_\lambda = I_{\lambda0}\,e^{-a_\lambda p x} \tag{13.29}$$

In terms of the transmissivity τ_λ, $I_\lambda = I_{\lambda0}\tau_\lambda$, and hence the transmissivity is given by

$$\tau_\lambda = e^{-a_\lambda p x} \tag{13.30}$$

Since $\tau_\lambda + \alpha_\lambda = 1$ for a gas, it follows that

$$\alpha_\lambda = 1 - e^{-a_\lambda p x} \tag{13.31}$$

This is also equal to the emissivity ε_λ if Kirchhoff's law is assumed to be valid.

In order to calculate a radiation exchange between a gas volume and an enclosing surface, account must be taken of radiation entering the gas volume from all directions. This would lead to the determination of emissivity (ε_g) and absorptivity (α_g) for a particular size and shape of gas volume. Such calculations are necessarily complex. However Hottel[9] has shown that gas volume shapes of practical interest can be compared to equivalent hemispheres, where the radiation from the surface to the centre of the base has a constant path length, the radius, which is known as the mean beam length for gas volume. Table 13.1 shows examples of mean beam length for simple shapes.

Table 13.1. Beam lengths of some simple gas volume shapes

Shape	Characteristic dimension, D	Mean beam length
Sphere	Diameter	$0.66 D$
Infinite Cylinder	Diameter	D
Cube	Length of side	$0.66 D$
Space outside infinite bank of tubes, centres on equilateral triangles, diameter = clearance	Clearance	$3.4 D$

The mean beam length of any shape may be calculated approximately as $L = 3.4 \times$ (volume of gas)/(area of enclosing surface).

Hottel has also made available extensive empirical data on the emissivities of radiating gases which are presented as a function of the product of partial pressure and beam length. Data for carbon dioxide and water vapour are given in Appendix 4, together with the procedure for calculating ε_g and α_g for gas mixtures.

13.8.1 Calculation of Radiation Exchange between Non-luminous Gases and Containing Surfaces

(a) *Black Surfaces.* If the gas volume is enclosed by a black surface of area A, the rate of radiation from the gas to the surface is

$$Q_{(g-s)} = \varepsilon_g A \sigma T_g^4 \tag{13.32}$$

where T_g is the absolute temperature of the gas. The rate of radiation from the surface absorbed by the gas is

$$Q_{(s-g)} = \alpha_g A \sigma T_g^4$$

Hence the net exchange between the gas and surface is

$$Q_{(gs)} = \sigma A(\varepsilon_g T_g^4 - \alpha_g T_s^4) \tag{13.33}$$

(b) *Grey Surfaces.* For the case of grey surfaces, the electrical network analogy of Section 13.7 may be used. Figure 13.16 shows a gas contained by two grey surfaces at T_1 and T_2 together with the analogy circuit. It is assumed all radiation leaving surface 1 is transmitted through the gas to reach surface 2.

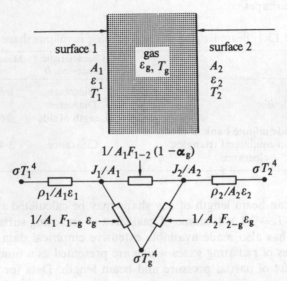

Fig. 13.16. *Electrical analogy of gas radiation involving two grey surfaces.*

The radiation leaving surface 1 transmitted to surface 2 is $J_1 F_{1-2} \tau_g$, and that leaving surface 2 transmitted to surface 1 is $J_2 F_{2-1} \tau_g$. Hence the net exchange between surfaces by transmission is $J_1 F_{1-2} \tau_g - J_2 F_{2-1} \tau_g$. Using the reciprocal relationship $A_1 F_{1-2} = A_2 F_{2-1}$,

$$Q_{(s_1 s_2)} = A_1 F_{1-2}(1 - \alpha_g)(J_1/A_1 - J_2/A_2)$$

This gives the top resistance in the circuit between the J_1/A_1 and J_2/A_2 nodes. The gas emits $\varepsilon_g \sigma T_g^4$, of this $F_{g-1} \varepsilon_g \sigma T_g^4$ reaches surface 1.

The radiation leaving surface 1 absorbed by the gas is $J_1 F_{1-g}\alpha_g$, hence the exchange between surface 1 and the gas is

$$Q_{(s_1g)} = A_1 F_{1-g}\varepsilon_g (J_1/A_1 - \sigma T_g^4)$$

assuming $\alpha_g = \varepsilon_g$, and the reciprocal relation $A_1 F_{1-g} = A_g F_{g-1}$. A similar relation exists between the gas and surface 2. Thus the resistances between the J_1/A_1 and σT_g^4 nodes, and the J_2/A_2 and σT_g^4 nodes are obtained. The σT_g^4 node is a source or sink node unless there is no chemical reaction when it becomes a floating potential between the J_1/A_1 and J_2/A_2 values. For a gas contained by a single grey surface the circuit consists of the two resistances between the σT_1^4, J_1/A_1 and σT_g^4 nodes only.

13.9 Solar Radiation

Interest in solar radiation as an energy source has increased substantially in recent years as the finite limits of fossil fuel reserves have become more apparent. Even in Great Britain useful amounts of sunshine are available either for direct conversion to electricity in photo-voltaic cells or for absorption as low grade heat in flat plate collectors.

The energy reaching the outer atmosphere of the earth may be calculated approximately by assuming the earth to be seen as a disc of radius 6436 km at a radial distance of $150\cdot6 \times 10^6$ km from the sun. Hence the percentage of the sun's radiation reaching the earth is $(\pi \times 6436^2)/(4\pi \times 10^{12} \times 150\cdot6^2) \times 100 = 4\cdot56 \times 10^{-8}$ per cent. The sun may be assumed a black-body emitter having a surface temperature of 6000 K and surface area of $6\cdot131 \times 10^{18}$ m^2, therefore the emission is $56\cdot7 \times 10^{-12} \times 6\cdot131 \times 10^{18} \times 6000^4 = 4\cdot506 \times 10^{23}$ kW. Consequently the amount reaching the outer atmosphere based on the earth's disc area is $21\cdot4 \times 10^{13}$ kW, or $1\cdot646$ kW/m^2 of earth's surface.

The measured quantity of energy received on unit area perpendicular to the sun's rays outside the earth's atmosphere at the mean distance between the sun and the earth is known as the solar constant. This is $2\cdot00$ cal/(cm^2 min) or $1\cdot388$ kW/m^2, rather less than the result of the approximate calculation.

Amounts reaching the surface of the earth which may be put to use are very much less, and indeed, at Kew the annual mean radiation received averaged over 24 hours is of the order of 100 W/m². Corresponding figures for the United States and Australia are about 180 and 200 W/m².

From these figures it is possible to carry out feasibility studies of, say, flat plate collectors for domestic hot water and lighting, bearing in mind a collection efficiency of 10–15 per cent for conversion to electricity using semi-conductor devices, and 50–70 per cent for conversion to low-grade heat, by absorption on black surfaces under glass.

EXAMPLE 13.4

A flat plate solar collector has a selective surface with an absorptivity of 0·92 and an emissivity of 0·15. The coefficient for heat loss by convection is 0·003 kW/(m² K). If the area is 20 m² calculate the rate of energy collection, and the collection efficiency at a time when the solar insolation is 800 W/m², for a collector temperature of 60°C and an ambient temperature of 18°C.

Solution. The rate of collection is $0·92 \times 0·8$ kW, less the losses, per m². The radiation loss is $\sigma \varepsilon T^4$ per m² $= 56·7 \times 10^{-12} \times 0·15 \times 333^4 = 0·105$ kW/m². The convection loss is $0·003 (60 - 18) = 0·126$ kW/m².

Hence the net rate of collection is $0·92 \times 0·8 - 0·105 - 0·126 = 0·505$ kW/m². Hence the rate of collection for 20 m² $= 10·1$ kW. The collection efficiency $= (0·505/0·8) \times 100 = 63·1$ per cent.

PROBLEMS

1. An electric oven is internally a 0·3 m cube. The back wall is maintained at 300°C and other surfaces may be assumed to be at a uniform 100°C. The emissivity of all surfaces is 0·8. Calculate the grey-body configuration factor and the net radiation transfer from the heated wall. (Ans.: 0·77; 0·348 kW.)

2. A spherical thermocouple is situated at the middle of the length and on the axis of a pipe of length L and diameter D to measure the temperature of the gas flowing through the pipe. Assuming that the couple is so small in comparison with the duct that the surface of the couple is always perpendicular to the direction of radiation, deduce from first principles the expression

$$F_{h-c} = \frac{L}{\sqrt{(D^2 + L^2)}}$$

which gives the 'area factor' of the system. Both the couple and the pipe may be considered to be black bodies.

In an installation similar to that described above, the thermocouple is 3 mm in diameter and the pipe is 0·92 m long and 0·31 m in diameter. The gas temperature is 149°C and the internal surface of the pipe is 65·6°C. Heat is being transferred from the gas to the couple at the rate of 0·79 kW/m². What would be the reading of the thermocouple if both the couple and the pipe can be considered as black? The expression for the area factor given above may be used. (Ans. 136°C.) (*King's College, London*).

3. A small disc of area A_1 is concentric and parallel with a large disc of area A_2 and diameter D, and is spaced at a distance R from it. Show that for the discs:

$$F_{1-2} = \frac{D^2}{4R^2 + D^2}$$

A furnace chamber is a cylinder 2 m long by 1 m diameter. At one end there is a window 20 mm diameter on the axis of the cylinder. A pyranometer with an element temperature of 50°C placed at the window receives a radiation transfer of 1·97 W from the opposite circular end. Calculate the temperature of that end. (Ans. 900°C.)

4. Show that for a grey body of area A_1 and emissivity ϵ_1 enclosed by a surface of area A_2 and emissivity ϵ_2 the grey body configuration factor is given by:

$$\mathscr{F}_{1-2} = \frac{1}{1/\epsilon_1 + (A_1/A_2)(1/\epsilon_{2-1})}$$

which may be approximated to $\mathscr{F}_{1-2} = \epsilon_1$ if A_2 is large in comparison with A_1.

A billet of surface area 1·8 m² is heated in a rectangular furnace 3 m by 3·5 m by 2·5 m, the whole surface of which is radiating. For $\epsilon_1 = 0·8$ and $\epsilon_2 = 0·7$, calculate the percentage error in using the approximate result for $\mathscr{F}_{1-2}$ in calculating the radiation exchange. (Ans. 1·05 per cent.)

5. Two large parallel plates, 1 and 2, having emissivities on their inner faces of 0·5 and 0·8 are maintained at 300° and 100°C respectively. A third plate having unknown emissivities on its faces A and B is placed between the other two plates. When face A is pointing towards plate 1, the third plate reaches an equilibrium of 278°C. When the third plate is turned round so that face B is pointing towards plate 1, its equilibrium temperature drops to 140°C. Determine the emissivities of the two faces A and B. (Ans. Face A, $\varepsilon = 0·916$, Face B, $\varepsilon = 0·102$.) (*The City University*).

6. A plane surface of emissivity ϵ_1 and at temperature T_1 radiates to a parallel shield at temperature T_2. The shield has an emissivity ϵ_2 facing the plane surface, and an emissivity ϵ_3 facing the surroundings at T_3, which may be assumed large. Show that the radiation from the plane surface to the surroundings, via the shield, per unit area, is given by:

$$Q_{1-3} = \frac{\sigma(T_1^4 - T_3^4)}{1/\epsilon_1 + 1/\epsilon_2 + 1/\epsilon_3 - 1}$$

For the situation described $T_1 = 500°C$, $\epsilon_1 = 0.4$, $\epsilon_2 = 0.2$, $\epsilon_3 = 0.05$, and $T_3 = 30°C$. Calculate the radiation from the plane surface to the surroundings per unit area, and calculate the temperature of the radiation shield. (Ans. 742 W/m², 449°C.)

7. Calculate the rate of energy absorption on a flat plate heat collector positioned normally to the sun's radiation, given the information in Section 14·9, plus the following: collector area 15 m², absorptivity to solar radiation 0·96, emissivity of plate 0·2, transmissivity of upper atmosphere 0·626, collector surface temperature 68°C, atmospheric temperature 20°C, natural convection coefficient from collector 0·003 kW/(m² K). (Ans. 10·35 kW.)

8. A billet reheating furnace has a brick interior 10 m × 3 m × 1 m. The brick surface has an emissivity of 0·85 and is maintained at 900°C. Billets pass slowly and continuously through the furnace on a moving floor and may be assumed to receive heat transfer to an exposed surface area of 30 m². Combustion gases, assumed transparent to radiation, at 900°C pass through the furnace. The convection coefficient between gases and billets is 0·05 kW/(m² K). The emissivity of the billets is 0·7. Calculate the total furnace heat output for a mean billet temperature of 450°C, and the percentages of this output which are due to (a) radiation, and (b) convection. (Ans. 2485 kW, (a) 72·8%, (b) 27·2%.) (*The City University*).

9. A rectangular furnace chamber has a floor of 5 m by 3·5 m. The height is 4 m and the upper 1·5 m of all four vertical surfaces are covered in radiant heaters having a temperature of 600°C. These heaters radiate to the floor at 150°C. By considering off-set perpendicular rectangles, calculate the configuration factor between the whole area and the floor. Assuming the surfaces are black, calculate the radiation rate to the floor. (Ans. 0·126, 99·67 kW.)

10. An annular combustion chamber is contained between an inner cylinder of 1·83 and an outer cylinder of 2·44 m diameter. Combustion gases within the annular space have a mean temperature of 870°C and they con-

tain 8 per cent by volume of each of carbon dioxide and water vapour. The pressure within the combustion chamber is 2 atmospheres. The outer cylinder is maintained at 424°C and the inner cylinder at 488°C. Calculate the net radiation from the gas per m length of chamber, and the heat extracted from each wall. The emissivity of the inner wall is 0·9, and of the outer wall 0·6. (Ans. 311 kW/m, inner 155·9 kW/m, outer 154·9 kW/m.) (*The City University*).

11. A room having a floor of 5 m by 4 m and a height of 3 m is heated by underfloor heating elements, so that the floor temperature is 35°C. The wall and ceiling temperature is 18°C. The floor emissivity is 0·8 and for the walls and ceiling the emissivity is 0·7. The room is heated by radiation and convection. For natural convection it may be assumed that $Nu = 0·14(GrPr)^{0.33}$ with the appropriate length being the floor mean side of 4·5 m. Calculate the total energy input to the floor heaters assuming this is dissipated entirely by radiation and convection to the room. (Ans. 2179·68 W, giving 1517·92 W radiation and 661·76 W convection.)

REFERENCES

1. Jakob, M. *Heat Transfer*, Vol. 1, John Wiley and Sons, Inc., New York (1949).
2. Planck, M. *The Theory of Heat Radiation* (Translation) Dover (1959).
3. Kirchhoff, G. *Ostwalds Klassiker d. exakten Wissens.*, 100, Leipzig (1898).
4. Lambert, J. H. *Photometria* (1860).
5. Sparrow, E. M. and Cess, R. D. *Radiation Heat Transfer*. Brooks/Cole Publishing Company (1970).
6. Chapman, A. J. *Heat Transfer*, 3rd ed. The Macmillan Company, New York (1974).
7. Oppenheim, A. K. *Amer. Soc. Mech. Engs*, Paper 54-A75 (1954).
8. Hottel, H. C. *Notes on Radiant Heat Transmission*, Chem. Eng. Dept., M.I.T. (1951).
9. McAdams, W. H. *Heat Transmission* (Chapter 4, Hottel, H. C.), 3rd ed., McGraw-Hill Book Company, Inc., New York (1954).

Appendix 1: Heat Transfer Literature

The following is a list of journals, proceedings, and bibliography which may be consulted in order to keep abreast of the most recently published work in heat transfer.

The International Journal of Heat and Mass Transfer, Pergamon Press, monthly

The Journal of Mechanical Engineering Science, The Institution of Mechnical Engineers, bi-monthly.

Journal of Heat Transfer, Transactions of the American Society of Mechanical Engineers, Series C, quarterly.

Proceedings of the International Heat Transfer Conferences, e.g., 4th 1970 (Paris), 5th 1974 (Tokyo), Elsevier Publishing Company, Amsterdam.

Progress in Heat and Mass Transfer, Monograph Series of the International Journal of Heat and Mass Transfer, Pergamon Press.

Advances in Heat Transfer, Academic Press, New York.

Proceedings of the Heat Transfer and Fluid Mechanics Institute, Stanford University Press, California.

Heat Bibliography, HMSO London, annual.

Reports of the National Engineering Laboratory, East Kilbride, (available on request).

The Engineering Index, Engineering Index, Inc., New York.

Applied Science and Technology Index, The H. W. Wilson Company, New York.

The British Technology Index, The Library Association, London.

ISMEC Bulletin, Information Service in Mechanical Engineering. The Institution of Mechanical Engineers.

Science Abstracts A, Physics Abstracts, The Institution of Electrical Engineers.

Science Abstracts B, Electrical and Electronic Abstracts, The Institution of Electrical Engineers and The Institute of Electrical and Electronic Engineers, Inc.

236

Appendix 2: Units and Conversion Factors

SI units are used exclusively in this book. However, much of the existing heat transfer literature is in British units, and SI–British conversion factors are therefore included. The kJ and kW are accepted alternatives to the J and W in the use of SI units. They are the units of energy and power generally used in the teaching of engineering thermodynamics and are the preferred units used in this book. For a complete discussion see *The Use of SI Units*, published by the British Standards Institution, PD 5686: 1972.

The Basic SI units are:

Mass $1\ \text{kg} = 2 \cdot 2046\ \text{lb}$
Length $1\ \text{m} = 3 \cdot 2808\ \text{ft}$
Time $1\ \text{s} = 2 \cdot 778 \times 10^{-4}\ \text{h}$
Temperature $1\ \text{K} = 1 \cdot 8\ °\text{Rankine}$

Derived SI units are:

Force $1\ \text{N} = 0 \cdot 2248\ \text{lbf}$ (1 newton $= 1\ \text{kg m/s}^2$)
Pressure $1\ \text{Pa} = 14 \cdot 5 \times 10^{-5}\ \text{lbf/in}^2$ (1 pascal $= 1\ \text{N/m}^2$)
               $1\ \text{bar} = 10^5\ \text{Pa} = 14 \cdot 5\ \text{lbf/in}^2$
Density $1\ \text{kg/m}^3 = 0 \cdot 06243\ \text{lb/ft}^3$
Specific
   volume $1\ \text{m}^3/\text{kg} = 16 \cdot 0179\ \text{ft}^3/\text{lb}$
Energy $1\ \text{J} = 1\ \text{Nm}; 1\ \text{kJ} = 10^3\ \text{Nm} = 737 \cdot 6\ \text{ft lbf}$
Power $1\ \text{W} = 1\ \text{N m/s}; 1\ \text{kW} = 737 \cdot 6\ \text{ft lbf/s} = 1 \cdot 341\ \text{h.p.}$

Conversion Factors for Heat Transfer Units

Physical quantity	SI	British units	Conversion factor*	Reciprocal conversion factor*
Q	kW	Btu/h	$2{\cdot}931 \times 10^{-4}$	$3{\cdot}412 \times 10^{3}$
q	kW/m²	Btu/(ft² h)	$3{\cdot}155 \times 10^{-3}$	$3{\cdot}170 \times 10^{2}$
h, U	kW/(m² K)	Btu/(ft² h °F)	$5{\cdot}678 \times 10^{-3}$	$1{\cdot}761 \times 10^{2}$
k	kW/(m K)	Btu/(ft h °F)	$1{\cdot}731 \times 10^{-3}$	$5{\cdot}777 \times 10^{2}$
c_p	kJ/(kg K)	Btu/(lb °F)	$4{\cdot}1868$	$0{\cdot}2388$
μ	Pa s	lb/(ft h)	$4{\cdot}134 \times 10^{-4}$	$2{\cdot}419 \times 10^{3}$
	(Note: 1 Pa s = 10 dyn s/cm² = 10 poise)			
$\nu, \alpha, \varepsilon, D$	m²/s	ft²/h	$2{\cdot}581 \times 10^{-5}$	$3{\cdot}874 \times 10^{4}$
	(Note: 1 m²/s = 10⁴ cm²/s = 10⁴ stokes, unit of dynamic viscosity)			
τ, P, p	Pa	lbf/ft²	$9{\cdot}931 \times 10^{5}$	$1{\cdot}007 \times 10^{-6}$
τ, P, p	Pa	lbf/in²	$6{\cdot}897 \times 10^{3}$	$1{\cdot}450 \times 10^{-4}$

* Multiply the numerical value in British units by the conversion factor to obtain the equivalent in SI; multiply the numerical value in SI by the reciprocal conversion factor to obtain the equivalent in British units.

Appendix 3: Tables of Property Values

Table A.1. Thermal Properties of Solids: Metals

	Properties at 20°C				$k \times 10^3$, kW/(m K)				
	ρ $\left(\dfrac{kg}{m^3}\right)$	$c_p \times 10^3$ $\left(\dfrac{kJ}{kg\,K}\right)$	$k \times 10^3$ $\left(\dfrac{kW}{m\,K}\right)$	α $\left(\dfrac{m^2}{s}\right)$	100	200	300 (°C)	400	600
Aluminium, pure	2707	896	204.	$8{\cdot}42 \times 10^{-5}$	206	215	229	249	
Duralumin, 94–96 Al, 3–5 Cu	2787	883	164	6·68	182	194			
Lead	11,370	130	34·6	2·34	33·4	31·5	29·8		
Iron, pure	7897	452	72·7	2·03	67·5	62·3	55·4	48·5	39·8
Iron, wrought, C < 0·5%	7849	460	58·9	1·63	57·1	51·9	48·5	45·0	36·4
Iron, cast, C ≈ 4%	7272	419	51·9	1·70					
Carbon steel, C ≈ 0·5%	7833	465	53·7	1·47	51·9	48·5	45·0	41·5	34·6
Carbon steel, C = 1·5%	7753	486	36·4	0·97	36·3	36·3	34·6	32·9	31·2
Nickel steel, 10%	7945	460	26·0	0·72					
Nickel steel, 30%	8073	460	12·1	0·33					
Nickel steel, 50%	8266	460	13·8	0·36					
Nickel steel, 70%	8506	460	26·0	0·67					
Nickel steel, 90%	8762	460	46·7	1·16					
Chrome steel, 1%	7865	460	60·6	1·67	55·4	51·9	46·7	41·5	36·4
Chrome steel, 5%	7833	460	39·8	1·11	38·1	36·4	36·4	32·9	29·4
Chrome steel, 10%	7785	460	31·2	0·87	31·2	31·2	29·4	29·4	31·2
Cr–Ni steel, 18% Cr, 8% Ni	7817	460	16·3	0·44	17·3	17·3	19·0	19·0	22·5
Ni–Cr steel, 20% Ni, 15% Cr	7865	460	14·0	0·39	15·1	15·1	16·3	17·3	19·0
Manganese steel, 2%	7865	460	38·1	1·05	36·4	36·4	36·4	34·6	32·9

239

Table A.1. *Continued*

	Properties at 20°C				$k \times 10^3$ kW/(m K)				
	ρ $\left(\dfrac{kg}{m^3}\right)$	$c_p \times 10^3$ $\left(\dfrac{kJ}{kg\,K}\right)$	$k \times 10^3$ $\left(\dfrac{kW}{m\,K}\right)$	α $\left(\dfrac{m^2}{s}\right)$	100	200	300 (°C)	400	600
Tungsten steel, 2%	7961	444	62.3	1.76×10^{-5}	58.9	53.7	48.5	45.0	36.4
Silicon steel, 2%	7673	460	31.2	0.89					
Copper, pure	8954	383	386	11.2	379	374	369	364	353
Bronze, 75 Cu, 25 Sn	8660	343	26.0	0.86					
Brass, 70 Cu, 30 Zn	8522	385	111	3.41	128	144	147	147	
German silver, 62 Cu 15 Ni, 22 Zn	8618	394	24.9	0.73	31.2	39.8	45.0	48.5	
Constantan, 60 Cu, 40 Ni	8922	410	22.7	0.61	22.2	26.0			
Magnesium, pure	1746	1013	171	9.71	168	163	158		
Molybdenum	10,220	251	123	4.79	118	114	111	109	106
Nickel, 99.9% pure	8906	446	90.0	2.27	83.1	72.7	64.0	58.9	
Silver, 99.9% pure	10,520	234	407	16.6	415	374	362	360	
Tungsten	19,350	134	163	6.27	151	142	133	126	113
Zinc, pure	7144	384	112	4.11	109	106	100	93.5	
Tin, pure	7304	227	64.0	3.88	58.9	57.1			

Adapted from Table A–1, E. R. G. Eckert and R. M. Drake, Jr., *Heat and Mass Transfer*, McGraw-Hill Book Company, New York (1959).

Table A.2. Thermal Properties of Solids: non-Metals

	$c_p \times 10^3$ $\left(\dfrac{kJ}{kg\,K}\right)$	ρ $\left(\dfrac{kg}{m^3}\right)$	t (°C)	$k \times 10^3$ $\left(\dfrac{kW}{m\,K}\right)$	α $\left(\dfrac{m^2}{s}\right)$
Bakelite	1590	1273	20	0·232	$0\cdot0114 \times 10^{-5}$
Bricks:					
Common	837	1602	20	0·692	0·0516
Face		2050	20	1·32	
Chrome	837	3011	200	2·32	0·0929
			550	2·48	0·0981
			900	1·99	0·0800
Diatomaceous earth			204	0·242	
(fired)			872	0·312	
Fire clay					
(burnt 1450°C)	963	2323	500	1·28	0·0568
			800	1·37	0·0619
			1100	1·402	0·0619
Magnesite	1130		204	3·81	
			648	2·77	
			1204	1·90	
Concrete	879	1906– 2307	20	0·814– 1·40	0·0490– 0·0697
Glass, plate	837	2707	20	0·762	0·0336
Plaster, gypsum	837	1442	21	0·485	0·0413
Stone:					
Granite	816	2643		1·73– 3·98	0·0800– 0·183
Limestone	908	2483	99	1·26	0·0568
			299	1·33	0·0594
Marble	808	2499– 2707	20	2·77	0·0394
Sandstone	712	2163– 2307	20	1·63– 2·08	0·106– 0·127
Wood, cross grain:					
Cypress		464	30	0·097	
Fir	2721	417	24	0·109	0·0095
Oak	2387	609–481	30	0·166	0·0126
Yellow pine	2805	641	24	0·147	0·0083
Wood, radial:					
Oak	2387	609– 481	20	0·173– 0·207	0·0111– 0·0121
Fir	2721	417	20	0·138	0·0124

Table A.2. *Continued*

	$c_p \times 10^3$ $\left(\dfrac{kJ}{kg\,K}\right)$	ρ $\left(\dfrac{kg}{m^3}\right)$	t (°C)	$k \times 10^3$ $\left(\dfrac{kW}{m\,K}\right)$	α $\left(\dfrac{m^2}{s}\right)$
Asbestos	816	577	0	0·151	
	816	577	100	0·192	
Cotton	1298	80·1	20	0·0589	0·194
Cork, board		160	30	0·0433	
Cork, expanded scrap	1884	44·8–119	20	0·0363	0·0155–0·0439
Earth, coarse gravelly	1842	2050	20	0·519	0·0139
Felt, wool		330	30	0·0519	
Fibre, insulating board		237	21	0·0485	
Glass wool	670	200	20	0·0398	0·0284
Ice	1926	913	0	2·22	0·124
Silk	1382	57·7	20	0·0363	0·0439

Adapted from A. J. Chapman, *Heat Transfer*, The Macmillan Company, New York (1960); L. S. Marks, *Mechanical Engineers' Handbook*, 5th ed., McGraw-Hill Book Company, Inc., New York (1951); W. H. McAdams, *Heat Transmission*, 3rd ed., McGraw-Hill Book Company, Inc., New York (1954); and E. R. G. Eckert and R. M. Drake, Jr., *Heat and Mass Transfer*, McGraw-Hill Book Company, Inc., New York (1959).

Table A3. Thermal Conductivity of Some Building Materials

	$\rho\,(\text{kg/m}^3)$	$k\,(\text{W/(mK)})$
Asbestos cement sheet	1520	0·29–0·43
Asbestos felt	144	0·078
Asbestos insulating board	720–900	0·11–0·21
Asphalt, roofing	1920	0·58
Brick, common, dry	1760	0·81
Brick, wet	2034	1·67
Chipboard	350–1360	0·07–0·21
Concrete, gravel 1:2:4	2240–2480	1·4
vermiculite aggregate	400–880	0·11–0·26
cellular	320–1600	0·08–0·65
Cork, granulated, raw	115	0·046
slab, raw	160	0·05
Fibreboard	280–420	0·05–0·08
Glass, window	2500	1·05
Glassfibre, mat	50	0·033
Hardboard	560	0·08
Plasterboard, gypsum	1120	0·16
Polystyrene, expanded board	15	0·037
Polyurethane foam	30	0·026
Polyvinyl chloride, rigid foam	25–80	0·035–0·041
Roofing felt	960–1120	0·19–0·20
Tiles, clay	1900	0·85
Tiles, concrete	2100	1·10
Tiles, PVC asbestos	2000	0·85
Urea formaldehyde foam	8–30	0·032–0·038
Vermiculite granules	100	0·065
Wilton carpet		0·058

Table A3. *Continued*

U values for Building structures, based on the difference between inside and outside environment temperatures, and for sheltered, normal and severe external exposure, in $W/(m^2 K)$.

	Sheltered	Normal	Severe
260 mm cavity wall, 105 mm inner and outer leaves, plus 16 mm lightplaster on inner face	1·3	1·3	1·3
220 mm solid wall, with 16 mm light plaster	1·8	1·9	2·0
335 mm solid wall, with 16 m light plaster	1·4	1·5	1·6
Pitched roof, tiles on battens with roofing felt, roof space, foil backed plasterboard ceiling	1·4	1·5	1·6
As above, plus 50 mm glass fibre loft insulation	0·49	0·5	0·51
Window, single glazing, 30% area due to wood frame	3·8	4·3	5·0
As above, double glazing	2·3	2·5	2·7

From the CIBS Guide Book A, The Chartered Institution of Building Services Engineers, London. The above U values and thermal conductivities are a brief extract only (used by permission of the Institution).

Table A.4. Physical Properties of some Common Low Melting Point Metals

	Melting point (°C)	Boiling point (°C)	Temp. (°C)	ρ $\left(\frac{kg}{m^3}\right)$	μ Pa s	$c_p \times 10^3$ $\left(\frac{kJ}{kg\,K}\right)$	$k \times 10^3$ $\left(\frac{kW}{mK}\right)$	Pr
Bismuth	272	1480	315	10,010	$1 \cdot 62 \times 10^{-3}$	144	16·4	0·014
			760	9467	0·79	164	15·6	0·0084
Lead	328	1738	371	10,540	2·40	159	16·1	0·024
			704	10,140	1·37	155	14·9	0·016
Lithium	179	1318	204	506	0·59	4187	38·1	0·065
			983	442	0·42	4187		
Mercury	−39	357	10	13,570	1·59	138	8·14	0·027
			315	12,850	0·87	134	14·0	0·0084
Potassium	64	760	149	807	0·37	796	45·0	0·0066
			704	674	0·13	754	33·1	0·0031
Sodium	97	884	204	902	0·43	1340	80·3	0·0072
			704	779	0·18	1256	59·7	0·0038
Sodium–Potassium, 22% Na	19	826	93·5	849	0·49	946	24·4	0·019
			760	690	0·16	883		
Sodium–Potassium, 56% Na	−11	795	93·5	887	0·58	1130	25·6	0·026
			760	740	0·16	1042	28·9	0·058
Lead–Bismuth, 44·5% Pb	125	1670	288	10,350	1·76	147	10·7	0·024
			649	9835	1·15			

Adapted from Table 16–1, J. G. Knudsen and D. L. Katz, *Fluid Dynamics and Heat Transfer*, McGraw-Hill Book Company, Inc., New York (1958).

Table A.5. Thermal Properties of Saturated Liquids

t (°C)	ρ (kg/m³)	$c_p \times 10^3$ kJ/(kg K)	ν (m²/s)	$k \times 10^3$ kW/(m K)	α (m²/s)	Pr	β (1/K)
				Water (H_2O)			
0	1002	4218	0.179×10^{-5}	0.552	13.1×10^{-8}	13.6	0.18×10^{-3}
20	1001	4182	0.101	0.597	14.3	7.02	
40	994.6	4178	0.0658	0.628	15.1	4.34	
60	985.4	4184	0.0477	0.651	15.5	3.02	
80	974.1	4196	0.0364	0.668	16.4	2.22	
100	960.6	4216	0.0294	0.680	16.8	1.74	
120	945.3	4250	0.0247	0.685	17.1	1.446	
140	928.3	4283	0.0214	0.684	17.2	1.241	
160	909.7	4342	0.0189	0.680	17.3	1.099	
180	889.0	4417	0.0173	0.675	17.2	1.004	
200	866.7	4505	0.0160	0.665	17.1	0.937	
220	842.4	4610	0.0149	0.653	16.8	0.891	
240	815.7	4756	0.0143	0.635	16.4	0.871	
260	785.9	4949	0.0137	0.611	15.6	0.874	
280	752.5	5208	0.0135	0.580	14.8	0.910	
300	714.3	5728	0.0135	0.540	13.2	1.019	

Methyl Chloride (CH₃Cl)

			$\times 10^{-5}$		$\times 10^{-8}$		2.63×10^{-3}
−50	1053	1476	0·0320	0·215	13·9	2·31	
−40	1033	1483	0·0318	0·209	13·7	2·32	
−30	1017	1492	0·0314	0·202	13·4	2·35	
−20	999·4	1504	0·0309	0·196	13·0	2·38	
−10	981·4	1519	0·0306	0·187	12·6	2·43	
0	962·4	1538	0·0302	0·178	12·1	2·49	
10	942·4	1560	0·0297	0·171	11·7	2·55	
20	923·3	1586	0·0292	0·163	11·1	2·63	
30	903·1	1616	0·0287	0·154	10·6	2·72	
40	883·1	1650	0·0281	0·144	9·96	2·83	
50	861·2	1689	0·0274	0·133	9·21	2·97	

Freon (CCl₂F₂)

			$\times 10^{-5}$		$\times 10^{-8}$	
−50	1547	875·0	0·0310	0·0675	5·01	6·2
−40	1519	884·7	0·0279	0·0692	5·13	5·4
−30	1490	895·6	0·0253	0·0692	5·26	4·8
−20	1461	907·3	0·0235	0·0710	5·39	4·4
−10	1430	920·3	0·0221	0·0727	5·50	4·0
0	1397	934·5	0·0214	0·0727	5·57	3·8
10	1364	949·6	0·0203	0·0727	5·60	3·6
20	1330	965·9	0·0198	0·0727	5·60	3·5
30	1295	983·5	0·0194	0·0710	5·60	3·5
40	1257	1002	0·0191	0·0692	5·55	3·5
50	1216	1022	0·0189	0·0675	5·44	3·5

Table A.5. *Continued*

t (°C)	ρ (kg/m³)	$c_p \times 10^3$ kJ/(kg K)	ν (m²/s)	$k \times 10^3$ kW/(m K)	α (m²/s)	Pr	β (1/K)
Glycerin ($C_3H_5(OH)_3$)							
0	1276	2261	$8{\cdot}31 \times 10^{-3}$	0·282	$9{\cdot}83 \times 10^{-8}$	$84{\cdot}7 \times 10^3$	
10	1270	2320	3·00	0·284	9·65	31·0	
20	1264	2387	1·17	0·286	9·47	12·5	$0{\cdot}504 \times 10^{-3}$
30	1258	2445	0·50	0·286	9·29	5·38	
40	1252	2512	0·22	0·286	9·13	2·45	
50	1245	2583	0·15	0·287	8·93	1·63	
Ethylene glycol ($C_2H_4(OH)_2$)							
0	1130	2294	$5{\cdot}75 \times 10^{-5}$	0·242	$9{\cdot}34 \times 10^{-8}$	615	
20	1117	2382	1·92	0·249	9·39	204	
40	1101	2474	0·869	0·256	9·39	93	$0{\cdot}648 \times 10^{-3}$
60	1088	2562	0·475	0·260	9·31	51	
80	1078	2650	0·298	0·261	9·21	32·4	
100	1059	2742	0·203	0·263	9·08	22·4	

Engine oil (unused)

T	ρ	cp	ν	k	α	Pr	β
0	899	1796	4.28×10^{-3}	0.147	9.11×10^{-8}	47,100	0.702×10^{-3}
20	888	1880	0.90	0.145	8.72	10,400	
40	876	1964	0.24	0.144	8.33	2870	
60	864	2047	0.0839	0.140	8.00	1050	
80	852	2131	0.0375	0.138	7.69	490	
100	840	2219	0.0203	0.137	7.38	276	
120	829	2307	0.0123	0.135	7.10	175	
140	817	2395	0.0080	0.133	6.86	116	
160	806	2483	0.0056	0.132	6.63	84	

Mercury (Hg)

T	ρ	cp	ν	k	α	Pr	β
0	13,630	140.3	0.0124×10^{-5}	8.21	430×10^{-8}	0.0288	1.82×10^{-4}
20	13,580	139.4	0.0114	8.69	461	0.0249	
50	13,510	138.6	0.0104	9.40	502	0.0207	
100	13,390	137.3	0.00928	10.5	571	0.0162	
150	13,260	136.5	0.00853	11.5	635	0.0134	
200	13,150	136.1	0.00802	12.3	691	0.0116	
250	13,030	135.7	0.00764	13.1	740	0.0103	
316	12,850	134.0	0.00673	14.0	815	0.0083	

Adapted from Table A–3, E. R. G. Eckert and R. M. Drake, Jr., *Heat and Mass Transfer*, McGraw-Hill Book Company, Inc., New York (1959).

Table A.6. Thermal Properties of Gases at Atmospheric Pressure

Air

T (°K)	ρ (kg/m³)	$c_p \times 10^3$ kJ/(kg K)	ν (m²/s)	$k \times 10^3$ kW/(m K)	α (m²/s)	μ Pa s	Pr
250	1·413	1005	$0·949 \times 10^{-5}$	0·0223	$1·32 \times 10^{-5}$	$1·60 \times 10^{-5}$	0·722
300	1·177	1006	1·57	0·0262	2·22	1·85	0·708
350	0·998	1009	2·08	0·0300	2·98	2·08	0·697
400	0·883	1014	2·59	0·0337	3·76	2·29	0·689
450	0·783	1021	2·89	0·0371	4·22	2·48	0·683
500	0·705	1030	3·79	0·0404	5·57	2·67	0·680
550	0·642	1039	4·43	0·0436	6·53	2·85	0·680
600	0·588	1055	5·13	0·0466	7·51	3·02	0·680
650	0·543	1063	5·85	0·0495	8·58	3·18	0·682
700	0·503	1075	6·63	0·0523	9·67	3·33	0·684
750	0·471	1086	7·39	0·0551	10·8	3·48	0·686
800	0·441	1098	8·23	0·0578	12·0	3·63	0·689
850	0·415	1110	9·07	0·0603	13·1	3·77	0·692
900	0·392	1121	9·93	0·0628	14·3	3·90	0·696
950	0·372	1132	10·8	0·0653	15·5	4·02	0·699
1000	0·352	1142	11·8	0·0675	16·8	4·15	0·702
1100	0·320	1161	13·7	0·0723	19·5	4·40	0·706
1200	0·295	1179	15·7	0·0763	22·0	4·63	0·714
1300	0·271	1197	17·9	0·0803	24·8	4·85	0·722

Hydrogen

250	0·0981	14,060	8·06 × 10⁻⁵	0·156	11·3 × 10⁻⁵	7·92 × 10⁻⁶	0·713
300	0·0819	14,320	10·9	0·182	15·5	8·96	0·706
350	0·0702	14,440	14·2	0·206	20·3	9·95	0·697
400	0·0614	14,490	17·7	0·229	25·7	10·9	0·690
450	0·0546	14,500	21·6	0·251	31·6	11·8	0·682
500	0·0492	14,510	25·7	0·272	38·2	12·6	0·675
550	0·0447	14,330	30·2	0·293	45·2	13·5	0·668
600	0·0408	14,540	35·0	0·315	53·1	14·3	0·664
650	0·0349	14,570	45·5	0·351	69·0	15·9	0·659
700	0·0306	14,680	56·9	0·384	85·6	17·4	0·664
750	0·0272	14,820	69·0	0·412	102	18·8	0·676
800	0·0245	14,970	82·2	0·440	120	20·2	0·686
850	0·0223	15,170	96·5	0·464	137	21·5	0·703

Oxygen

200	1·956	913·1	0·795 × 10⁻⁵	0·0182	1·02 × 10⁻⁵	14·9 × 10⁻⁶	0·745
250	1·562	915·6	1·144	0·0226	1·58	17·9	0·725
300	1·301	920·3	1·586	0·0267	2·24	20·6	0·709
350	1·113	929·0	2·080	0·0307	2·97	23·2	0·702
400	0·976	942·0	2·618	0·0346	3·77	25·5	0·695
450	0·868	956·7	3·199	0·0383	4·61	27·8	0·694
500	0·780	972·2	3·834	0·0417	5·50	29·9	0·697
550	0·710	988·1	4·505	0·0452	6·44	32·0	0·700
600	0·650	1004	5·214	0·0483	7·40	33·9	0·704

Table A.6. Continued

T (°K)	ρ (kg/m³)	$c_p \times 10^3$ kJ/(kg K)	ν (m²/s)	$k \times 10^3$ kW/(m K)	α (m²/s)	μ Pa s	Pr
				Nitrogen			
200	1·711	1043	$0·757 \times 10^{-5}$	0·0182	$1·02 \times 10^{-5}$	$12·9 \times 10^{-6}$	0·747
300	1·142	1041	1·563	0·0262	2·21	17·8	0·713
400	0·854	1046	2·574	0·0333	3·74	22·0	0·691
500	0·682	1056	3·766	0·0398	5·53	25·7	0·684
600	0·569	1076	5·119	0·0458	7·49	29·1	0·686
700	0·493	1097	6·512	0·0512	9·47	32·1	0·691
800	0·428	1123	8·145	0·0561	11·7	34·8	0·700
900	0·380	1146	9·106	0·0607	13·9	37·5	0·711
1000	0·341	1168	11·72	0·0648	16·3	40·0	0·724
1100	0·311	1186	13·60	0·0685	18·6	42·3	0·736
1200	0·285	1204	15·61	0·0719	20·9	44·5	0·748
				Carbon dioxide			
250	2·166	803·9	$0·581 \times 10^{-5}$	0·0129	$0·740 \times 10^{-5}$	$12·6 \times 10^{-6}$	0·793
300	1·797	870·9	0·832	0·0166	1·06	15·0	0·770
350	1·536	900·2	1·119	0·0205	1·48	17·2	0·755
400	1·342	942·0	1·439	0·0246	1·95	19·3	0·738
450	1·192	979·7	1·790	0·0290	2·48	21·3	0·721
500	1·073	1013	2·167	0·0335	3·08	23·3	0·702
550	0·974	1047	2·574	0·0382	3·75	25·1	0·685
600	0·894	1076	3·002	0·0431	4·48	26·8	0·668

Carbon monoxide

			$\times 10^{-5}$		$\times 10^{-5}$	$\times 10^{-6}$	
250	0·841	1043	1·128	0·0214	1·51	15·4	0·750
300	1·139	1042	1·567	0·0253	2·13	17·8	0·737
350	0·974	1043	2·062	0·0288	2·84	20·1	0·728
400	0·854	1048	2·599	0·0323	3·61	22·2	0·722
450	0·762	1055	3·188	0·0436	4·44	24·2	0·718
500	0·682	1063	3·819	0·0386	5·33	26·1	0·718
550	0·620	1076	4·496	0·0416	6·24	27·9	0·721
600	0·568	1088	5·206	0·0445	7·19	29·6	0·724

Water vapour

			$\times 10^{-4}$		$\times 10^{-5}$	$\times 10^{-6}$	
380	0·586	2060	0·216	0·0246	2·04	12·7	1·060
400	0·554	2014	0·242	0·0261	2·24	13·4	1·040
450	0·490	1980	0·311	0·0299	3·07	15·3	1·010
500	0·441	1985	0·386	0·0339	3·87	17·0	0·996
550	0·400	1997	0·470	0·0379	4·75	18·8	0·991
600	0·365	2026	0·566	0·0422	5·73	20·7	0·986
650	0·338	2056	0·664	0·0464	6·66	22·5	0·995
700	0·314	2085	0·772	0·0505	7·72	24·3	1·000
750	0·293	2119	0·888	0·0549	8·83	26·0	1·005
800	0·274	2152	1·020	0·0592	10·0	27·9	1·010
850	0·258	2186	1·152	0·0637	11·3	29·7	1·019

Adapted from Table A–4, E. R. G. Eckert and R. M. Drake, Jr., *Heat and Mass Transfer*, McGraw-Hill Book Company, Inc., New York (1959).
(*Note*: At pressures other than atmospheric, the density can be determined from the ideal gas equation, $\rho = p/RT$. Hence at any given temperature $\rho = \rho_0(p/p_0)$ where p_0 is atmospheric pressure and ρ_0 is given in the table. k, μ, and c_p may be assumed independent of pressure. ν and α are inversely proportional to the density; hence at a given temperature are inversely proportional to the pressure.)

Table A.7. Normal Total Emissivity of Various Surfaces

	Ref.	t (°C)	Emissivity
Aluminium:			
Highly polished plate, 98·3% pure	11	237–576	0·039–0·057
Rough polish	1	100	0·18
Commercial sheet	1	100	0·09
Heavily oxidized	2	93–505	0·20–0·31
Al-surfaced roofing	5	38	0·216
Brass:			
Highly polished, 73·2 Cu, 26·7 Zn	11	247–357	0·028–0·031
Polished	1	100	0·06
Rolled plate, natural surface	10	22	0·06
Chromium, polished	1	100	0·075
Copper:			
Carefully polished electrolytic copper	6	80	0·018
Polished	1	100	0·052
Molten	3	1076–1278	0·16–0·13
Iron and steel:			
Steel, polished	1	100	0·066
Iron, polished	12	427–1028	0·14–0·38
Cast iron, polished	9	200	0·21
Cast iron, newly turned	10	22	0·44
Wrought iron, highly polished	16	38–249	0·28
Iron plate, completely rusted	10	19	0·69
Sheet steel, shiny oxide layer	10	24	0·82
Steel plate, rough	5	38–372	0·94–0·97
Cast iron, molten	15	1300–1400	0·29
Steel, molten	7	1522–1650	0·43–0·40
Stainless steel, polished	1	100	0·074
Lead, grey oxidized	10	24	0·28
Magnesium oxide	8	278–827	0·55–0·20
Nichrome wire, bright	14	49–1000	0·65–0·79
Nickel-silver, polished	1	100	0·135
Platinum filament	4	27–1230	0·036–0·192
Silver, polished, pure	11	227–627	0·02–0·032
Tin, bright tinned iron	10	23	0·043, 0·064
Tungsten filament	18	3320	0·39
Zinc, galvanized sheet iron, fairly bright	10	28	0·23

Table A.7. *Continued*

	Ref.	t (°C)	Emissivity
Asbestos board	10	23	0·96
Brick:			
Red, rough	10	21	0·93
Building	14	1000	0·45
Fireclay	14	1000	0·75
Magnesite, refractory	14	1000	0·38
Candle soot	17	97–272	0·952
Lampblack, other blacks	14	50–1000	0·96
Graphite, pressed, filed surface	8	249–516	0·98
Concrete tiles	14	1000	0·63
Enamel, white fused, on iron	10	19	0·90
Glass, smooth	10	22	0·94
Oak, planed	10	21	0·90
Flat black lacquer	5	38–94	0·96–0·98
Oil paints, 16 different, all colours	13	100	0·92–0·96
Aluminium paints, various	13	100	0·27–0·67
Radiator paint, bronze	1	100	0·51
Paper, thin, pasted on blackened plate	10	19	0·92, 0·94
Plaster, rough lime	16	10–87	0·91
Roofing paper	10	21	0·91
Water (calculated from spectral data)		0–100	0·95–0·963

(*Note:* When temperatures and emissivities appear in pairs separated by dashes, they correspond; and linear interpolation is permissible.)
By courtesy of H. C. Hottel, from *Heat Transmission*, 3rd ed., by W. H. McAdams, McGraw-Hill Book Company, Inc., New York (1954).

REFERENCES

1. Barnes, B. T., Forsythe, W. E., and Adams, E. Q. *J. Opt. Soc. Amer.*, Vol. 37, 804 (1947).
2. Binkley, E. R., private communication (1933).
3. Burgess, G. K. *Natl. Bur. Stand.*, Bull. 6, Sci. paper 121, 111 (1909).
4. Davisson, C., and Weeks, J. R. Jr. *J. Opt. Soc. Amer.*, Vol. 8, 581 (1924).
5. Heilman, R. H. *Trans. ASME*, FSP 51, 287 (1929).
6. Hoffman, K. *Z. Physik*, Vol. 14, 310 (1923).
7. Knowles, D., and Sarjant, R. J. *J. Iron and Steel Inst.* (London), Vol. 155, 577 (1947).
8. Pirani, M. *J. Sci. Instrum.*, Vol. 16, 12 (1939).
9. Randolf, C. F., and Overhaltzer, M. J. *Phys. Rev.*, Vol. 2, 144 (1913).
10. Schmidt, E. *Gesundh-Ing.*, Beiheft 20, Reihe 1, 1–23 (1927).

11. Schmidt, H., and Furthman, E. *Mitt. Kaiser-Wilhelm-Inst. Eisenforsch. Dusseldorf, Abhandle.*, Vol. 109, 225 (1928).
12. Snell, F. D. *Ind. Eng. Chem.*, Vol. 29, 89 (1937).
13. Standard Oil Development Company, personal communication (1928).
14. Thring, M. W. *The Science.of Flames and Furnaces*, Chapman and Hall, London (1952).
15. Thwing, C. B. *Phys. Rev.*, Vol. 26, 190 (1908).
16. Wamsler, F. *Z. Ver. deut. Ing.*, Vol. 55, 599 (1911); *Mitt. Forsch.*, Vol. 98, 1 (1911).
17. Wenzl, M., and Morawe, F. *Stahl u. Eisen*, Vol. 47, 867 (1927).
18. Zwikker, C. *Arch. néerland. sci.*, Vol. 9, 207 (1925).

Appendix 4
Gas Emissivities

The curves in Figs. A1 and A2 give respectively emissivities of carbon dioxide and water vapour. In each case there are separate curves for constant values of the product of partial pressure and mean beam length. As the total pressure is increased, the lines of the CO_2 spectrum broaden, and a correction factor from Fig. A3 is applied for pressures other than 1 atmosphere. In the case of water vapour, the emissivity depends on the actual partial pressure and the total pressure as well as on the product of partial pressure and beam length.

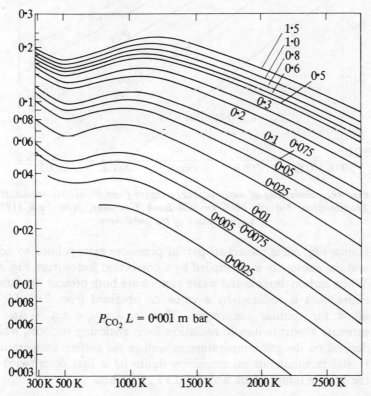

Fig. A1. Emissivity of carbon dioxide; adapted from W. H. McAdams Heat Transfer, McGraw-Hill Book Company, 3rd ed., New York (1954); by permission of the publishers.

257

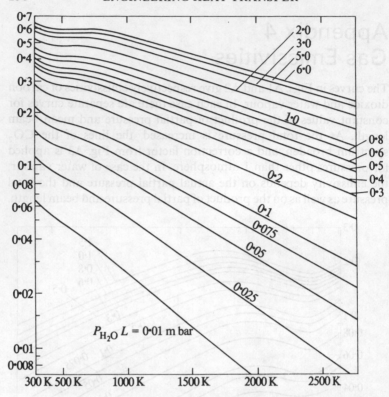

Fig. A2. Emissivity of water vapour; adapted from W. H. McAdams, Heat Transmission, 3rd ed. McGraw-Hill Book Company, New York (1954); by permission of the publishers.

Hence Fig. A2 is for actual partial pressures extrapolated to zero, and the emissivity is multiplied by a correction factor from Fig. A4. When carbon dioxide and water vapour are both present the sum of emissivities is reduced by a value $\Delta\varepsilon$ obtained from Fig. A5, to allow for mutual absorption. Thus $\varepsilon_g = \varepsilon_{H_2O} + \varepsilon_{CO_2} - \Delta\varepsilon$. To estimate absorptivities to radiation from enclosing surfaces, which depend on the gas temperature as well as the surface temperature, Hottel recommends an emissivity figure (ε) is first determined at the surface temperature and at $(pL)(T_s/T_g)$. Then

$$\alpha_{CO_2} = \varepsilon(T_g/T_s)^{0.65}$$

$$\alpha_{H_2O} = \varepsilon(T_g/T_s)^{0.45}$$

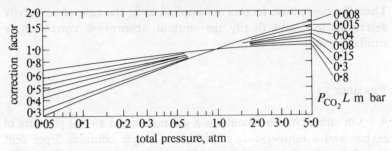

Fig. A3. Adapted from W. H. McAdams, Heat Transmission, 3rd ed., McGraw-Hill Book Company, New York (1954); by permission of the publishers.

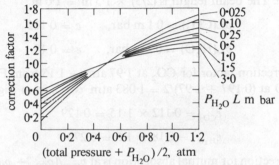

Fig. A4. Adapted from W. H. McAdams, Heat Transmission, 3rd ed., McGraw-Hill Book Company, New York (1954); by permission of the publishers.

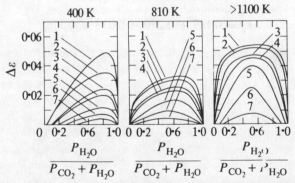

Fig. A5. Adapted from W. H. McAdams, Heat Transmission, McGraw-Hill Book Company, New York (1954); by permission of the publishers. For lines of constant $P_{CO_2} L + P_{H_2O} L$, in m bar, 1–1·5 m bar, 2–1·0 m bar, 3–0·6 m bar, 4–0·5 m bar, 6–0·2 m bar, 7–0·1 m bar.

Then the correction factors are applied as in the case of emissivity determination, and finally the mutual absorption correction is similarly made.

EXAMPLE

A 1·5 m cubic chamber contains a gas mixture at a total pressure of 2·0 bar and a temperature of 1000 K. The gas contains 5 per cent by volume of carbon dioxide and 10 per cent water vapour. Determine the emissivity of the gas mixture.

Solution. The beam length is $(2/3) \times 1·5$ m $= 1·0$ m.

$$pL(CO_2) = 0·1 \text{ m bar}, \qquad \varepsilon = 0·112$$

$$pL(H_2O) = 0·2 \text{ m bar}, \qquad \varepsilon = 0·18.$$

The correction factor for CO_2 at 1·97 atm $= 1·15$ from Fig. A3, and for H_2O at $(0·197 + 1·97)/2 = 1·083$ atm, is 1·5, from Fig. A4

$$\varepsilon_{CO_2} = 0·112 \times 1·15 = 0·129$$

$$\varepsilon_{H_2O} = 0·18 \times 1·5 = 0·270$$

The correction for mutual absorption is at $p_{H_2O}/(p_{CO_2} + p_{H_2O}) = 0·66$, and $pL(CO_2) + pL(H_2O) = 0·3$ m bar. From the set of curves at 1100 K, $\Delta\varepsilon = 0·035$, at 810 K, $= 0·016$. Hence $\Delta\varepsilon$ may be taken as 0·023.

$$\therefore \quad \varepsilon_g = 0·129 + 0·270 - 0·023 = 0·376$$

Index

266   INDEX